W9-AAB-219

MANUFACTURING STRATEGY

MANUFACTURING STRATEGY

How to Formulate and Implement a Winning Plan

JOHN MILTENBURG

Productivity Press
Portland, Oregon

 Additional copies of this book are available from the publisher. Discounts are available for multiple copies through the Sales Department (800-394-6868). Address all other inquiries to:

Productivity Press
P.O. Box 13390
Portland, OR 97213-0390
United States of America
Telephone: 503-235-0600
Telefax: 503-235-0909
E-mail: service@ppress.com

ISBN: 1-56327-071-4

Cover design by William Stanton
Page design and composition by Rohani Design, Edmonds, WA
Printed and bound by BookCrafters in the United States of America

Library of Congress Cataloging-in-Publication Data

Miltenburg, John.
Manufacturing strategy : how to formulate and implement a winning plan / John Miltenburg.
p. cm.
Includes bibliographical references and index.
ISBN 1-56327-071-4
1. Production planning. 2. Production control. I. Title.
TS176.M5429 1995
658.5—dc20 95-6352
CIP

02 01 00 99 98 97 10 9 8 7 6 5 4 3 2

TO MARGARET AND CORNELIS

CONTENTS

PUBLISHER'S MESSAGE

If you are a manager in manufacturing who hopes to take your company to the forefront of twenty-first century competition, you need this book. In *Manufacturing Strategy,* John Miltenburg offers a sensible and systematic method for evaluating your current production system in terms of the outputs it provides, and for planning the appropriate production strategy to be first in your market. As one advance reader has told us, "it's like an MBA in a book"—with a foundation firmly rooted in what has and hasn't worked in the field.

Part I of the book outlines the basic elements of a manufacturing system, starting with the six major manufacturing outputs—the factors customers consider in their purchasing decisions. Fast and reliable delivery, low cost, and consistent quality are the traditional set of outputs, but they aren't enough for today's markets. Many customers are also seeking performance—not just meeting specified, warranted quality, but surpassing it with advanced product features. In some markets, flexibility is the main event, the ability to change models, quantities, or other product attributes on request. Finally, there are markets that place the most value on innovativeness, the ability to provide unique features or products no one else offers.

Determining which outputs are most important to your business will point you to certain production systems that are best suited to produce those results. But success with a particular system usually depends not only on proper adjustment of the levers, but also on an increase in overall capability in each area. The JIT system, with its kanban coordination and attention to zero defects and zero inventory, is a case in point. It requires not just adjustment of the levers, but quantum improvement in capabilities. Part II of the book describes process improvement strategies that enable manufacturing organizations to increase capability to world class levels.

Each chapter in Part III focuses on a specific production system, treating in more depth the market outputs linked to each one, the lever adjustments required for success, and situational case studies of the manufacturing strategy system at work.

Miltenburg introduces the seven basic production systems and the outputs each system is best able to provide. The book describes each system in depth, from the customized job shop through line flow and just-in-time, to highly automated continuous flow systems. Each system handles different volumes and variety of products, and each requires a certain type of layout and material flow. *Manufacturing Strategy* arrays these variables in a worksheet matrix that makes it easy to see the interrelationships, and presents numerous situational examples of companies using different production systems.

Next, the book describes the various levers that make up a manufacturing system—human resources, organizational structure, production planning and control mechanisms, sourcing decisions, process technology, and facilities. The pattern of these six interacting levers determines which type of production system can operate. Thus, any changes to these basic elements should be made carefully to ensure that the resulting combination is in sync with the production system. Conversely, changing from one production system to another typically requires a change in one or more of these basic levers. Miltenburg gives three basic assessment questions every manufacturer should ask before making significant changes in any of these six levers.

Miltenburg then ties these elements together into a process for determining which is the optimal production system to use. He begins this with a competitive analysis, which looks at the target market in terms of desired outputs and their various attributes. The book uncovers a basic truth: No company can provide all six outputs at world class levels. There are tradeoffs between factors such as delivery and innovativeness, cost and flexibility. A competitive company strives to provide the strongest possible results in the outputs it must have to qualify for its market, and concentrates its effort on the one or two outputs that will give it an order-winning edge over other suppliers.

This book supplies a comprehensive, systems perspective for business analysis of the manufacturing process. Everyone contemplating a change in manufacturing systems, particularly a change to JIT, should study this book before starting out.

We would like to express our appreciation to John Miltenburg for publishing with us and in particular for his effort in preparing the illustrations. Thanks also to Diane Asay and Cheryl Berling Rosen, acquisitions; Karen Jones and Vivina Ree, editorial; Marianne L'Abbate, copyediting; Bill Stanton, design and cover; Susan Swanson, production; Rohani Design, graphics and composition; Lynn Hutchinski, index.

Norman Bodek
Publisher

ACKNOWLEDGMENTS

Many friends and colleagues have contributed to this book through their encouragement and day-to-day discussions on the topics in general and portions of the text in particular. I am grateful for the advice, criticism, and suggestions of a large number of manufacturing managers, especially Ted Cambridge, John Coles, Gary Greer, Philip Kirby, John Lockyer, Robert Marshall, George Miltenburg, Walter Petryschuk, and Philip Walker. I am also thankful for the help of Professors Marvin Carlson, Scott Edgett, Itzhak Krinsky, Ali Montazemi, Randolph Ross, Marvin Ryder, Edward Silver, and Jacob Wijngaard. Special mention must also be made of the contributions of four authorities in the field, Robert Hayes, Yasuhiro Monden, Taiichi Ohno, and Steven Wheelwright, whose work is the foundation on which the results in this book are built. Finally I wish to thank Diane Asay and Karen Jones at Productivity Press for bringing this book to a successful completion.

INTRODUCTION

The business strategy of a firm is the sum of the individual strategies of its component functions—finance, manufacturing, marketing, product development, service, and so on. In a successful firm these strategies interlock to provide the maximum competitive advantage of which the firm is capable. No function is left out and no function dominates. However, in some firms the business strategy is dominated by nonmanufacturing functions, with the result being "thrown over the wall" to manufacturing. Manufacturing is uncomfortable with strategic planning and, in trying to be all things to all people, struggles. This book tries to change this.

In many organizations employees both inside and outside the manufacturing function realize that manufacturing is struggling to provide what the organization needs to be successful in a competitive marketplace. The reason is not that manufacturing's capabilities have diminished, but that customer expectations and the level of manufacturing capability of the organizations' competitors have increased. Getting manufacturing to meet market expectations can be accomplished only by realigning manufacturing, making improvements, and increasing manufacturing capabilities. But how to do this is not at all clear to manufacturing managers or to their staff advisers.

Manufacturing is complex. Large numbers of employees—skilled and unskilled, line and staff, flexible and inflexible—work there. Formal and informal systems, good and bad traditions, old and new cultures, all coexist. Production is sometimes low volumes of highly engineered, customized products, sometimes medium volumes of high-performance products with short product life cycles, and sometimes high-quality, low-cost, high-volume commodities. Production processes are as varied as the products they produce. In the last ten years, countless new techniques and technologies have appeared, each presented as the way to dramatically improve manufacturing capability. A manufacturing plan or strategy is needed to bring some structure into this complex environment.

Manufacturing strategy can be thought of as the underlying pattern in the sequence of decisions made by manufacturing over a long time. When a formal manufacturing strategy exists, decisions follow a neat, logical pattern. When there is no strategy, the pattern is erratic and unpredictable. The essence of manufacturing strategy is to formulate explicitly how manufacturing decisions will be made so that manufacturing will help the organization achieve a long-term advantage over its competitors.

Many managers are familiar with the distinction between efficiency and effectiveness. *Efficiency consists of doing things right,* whereas *effectiveness consists of doing the right things.* Manufacturing strategy focuses on effectiveness first, then on efficiency; that is, strategy seeks to ensure that 1) the right things are being done, and then 2) that the right things are done well. It is not unusual to find manufacturers, without a strategy to guide them, very efficiently doing the wrong things.

Any process for developing a manufacturing strategy should do the following:

- Take into account the requirements of customers.
- Take into account what competitors are doing.
- Consider manufacturing's current capabilities.
- Consider all options available to manufacturing.
- List the outputs that manufacturing will provide and specify, in detail, the optimal set of changes that must be made to accomplish them.

Although many different processes have been used to develop manufacturing strategy, they share enough common elements that a comprehensive, general-purpose process can be framed. Such a process is presented in this book. Other general processes for formulating manufacturing strategy have been developed. The best known are those of Fine and Hax (1985) and Hill (1989). More specialized processes have been documented at Digital (Moody 1990), IBM (Adesso 1985, Baumann 1988), and Hewlett-Packard (Beckman et al. 1990). The general process outlined in this book picks up where these processes stop. It reflects the most recent experiences of manufacturers all over the world, and the most recent research on the subject. (See, for example, Leong, Snyder, and Ward 1990; Hayes, Wheelwright, and Clark 1988; and Hayes and Wheelwright 1984). Many companies have used this process, all with much success. They have found that the process strikes the right balance between difficulty of application and usefulness of results. The process is not so complex as to be impractical, and not so simple that the results are of no use.

The process is equally applicable to large and small companies. Despite the considerable experience large companies have with strategic planning, most will find that the process presented in this book is much richer and provides more insights than their current processes. Small companies, which are often uncomfortable with any type of strategic planning, will find this process simple enough for them and, at the same time, rich enough that it can be used for all their strategic planning. Most manufacturers are small. In the United States there are 355,000 small manufacturers, each employing fewer than 500 employees. Together they employ 8 million people, or 95 percent of the U.S. manufacturing base.

Features of the General-Purpose Process for Developing Manufacturing Strategy

Feature 1: Automatic Generation of All Alternatives

The book outlines in operational detail a general-purpose process for formulating manufacturing strategy and developing an implementation plan for that strategy. The process automatically generates all the alternatives available to manufacturing and evaluates each in turn against the needs of the organization. This helps top management explore the implications of alternate strategies. Because no alternatives are overlooked, the outcome of

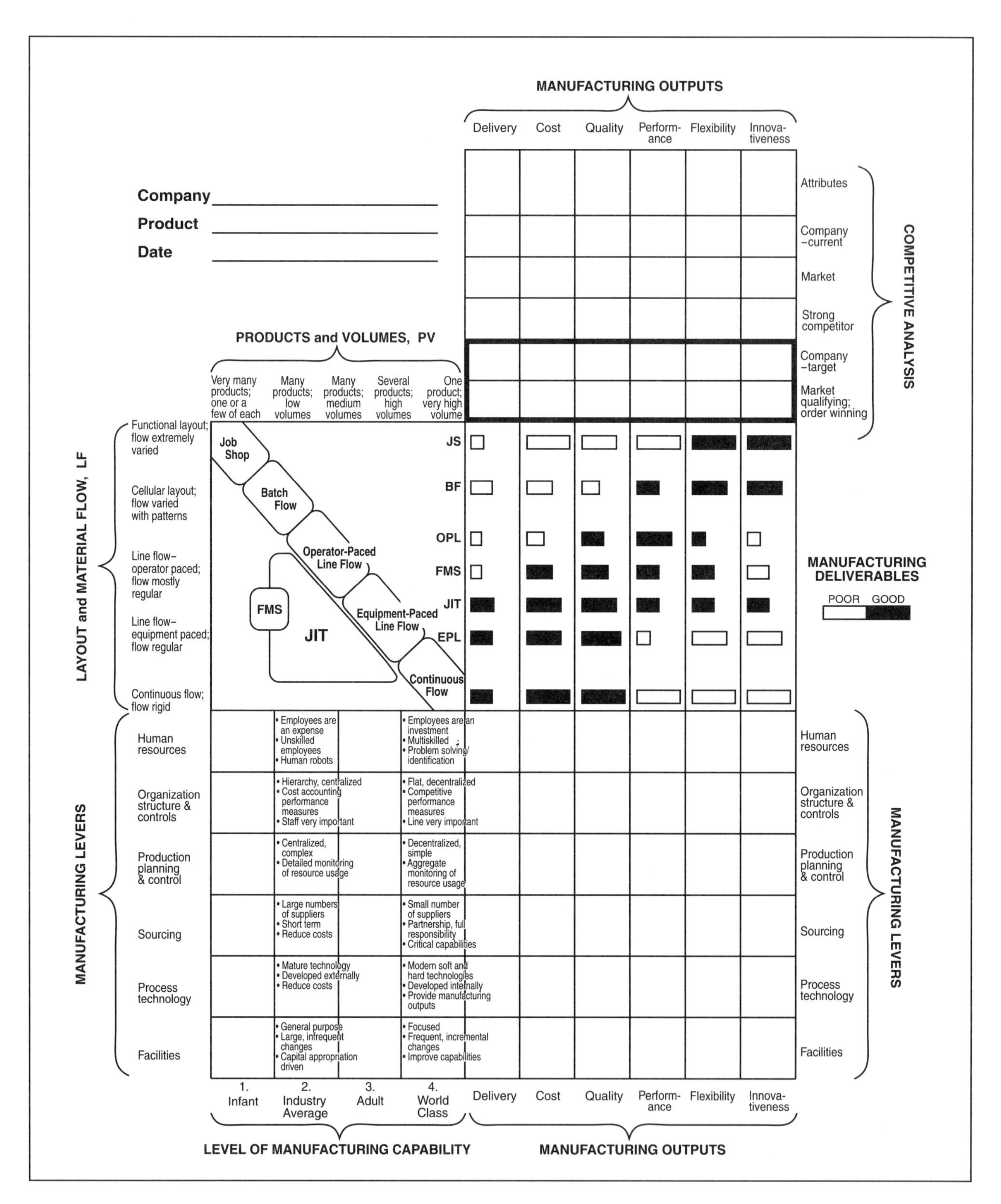

FIGURE A

The Manufacturing Strategy Worksheet

Company ____________ Production System ____________

Product ____________ Outputs – Market Qualifying ____________

Date ____________

– Order Winning ____________

	Current	Target

Manufacturing Levers	Elements	Projects: Set Course	Shoot and Aim	
Human resources	• • •	Top management awareness, acceptance of concepts, and commitment to execute		
Organization structure and controls	• • •			
Production planning and control	• • •			
Sourcing	• • •			
Process technology	• • •			
Facilities	• • •			
Targets and Deliverables		• Vision, leadership, visibility, support, active participation		
Time (Months)		0 3	9	12 18 24 30

FIGURE B

The Implementation Plan Worksheet

the process is the best possible manufacturing strategy for the organization. The process is summarized on two comprehensive, easy to use worksheets—the manufacturing strategy worksheet and the implementation plan worksheet. They are shown in Figures A and B.

FEATURE 2: CONDITIONS UNDER WHICH POPULAR CHANGES ARE APPROPRIATE

During the process of formulating a manufacturing strategy, three issues are addressed:

1. Where is manufacturing?
2. Where does manufacturing need to be?
3. What is the best way to move manufacturing from where it is to where it needs to be?

The answer to the last question is a list of changes that need to be made in manufacturing.

Many popular changes are discussed in the second part of this book. These include benchmarking, improvement approaches (such as total quality management, cycle time management, kaizen, and reengineering), focusing operations, soft technologies (such as concurrent engineering, small-lot production, supply line management, and total productive maintenance), and hard technologies (such as CNC, robots, and CAD/CAM). This book defines the strategic conditions in which these changes should be used. The result is more than "yes, it is appropriate" versus "no, it is not appropriate." The result is an understanding of the conditions where particular changes ought to be implemented, how changes should be grouped together, and how changes should be organized for implementation.

Feature 3: A Common Language

This book gives managers the skills they need to formulate and implement a manufacturing strategy that is optimal for their organization. An important consequence is the rich manufacturing strategy paradigm that is developed. Vocabulary, definitions, relationships, and so on, are developed for dealing with manufacturing problems at both strategic and operational levels. This paradigm facilitates discussion among manufacturing managers, many of whom have long been uncomfortable with strategic issues. It also makes communication clearer between manufacturing managers and those outside manufacturing, who will now have a better appreciation of what manufacturing can and cannot do.

How This Book Is Organized

This book is organized into three parts.

Part I

In the first part, a process is developed for formulating manufacturing strategy. The process consists of five elements:

1. Manufacturing outputs: What does manufacturing provide?
2. Production systems: How does manufacturing provide what is required?

3. Manufacturing levers: What parts of the production system should be adjusted?
4. Competitive analysis: What outputs should manufacturing provide and at what levels?
5. Manufacturing capability: Does manufacturing have sufficient capability for what is required?

Each element is sequentially discussed in the first five chapters. Chapters 6 and 7 synthesize these elements into a viable process for formulating manufacturing strategy. Each element interacts with every other element and their interactions are displayed on the worksheet in Figure A. Experience reveals that, of the many possible interactions among elements, only five or six will be important in a particular situation. The worksheet makes it easy to identify and analyze them.

Part II

The second part of the book presents changes and techniques frequently used in each of the five elements. For example, benchmarking (see Chapter 9) is a technique frequently used in the competitive analysis element and the manufacturing capability element. Figure C outlines the relationship among the techniques presented in the second part of the book and the elements in the manufacturing strategy process from the first part.

Part III

The third part of the book is an in-depth treatment of the seven different production systems that comprise the production systems element of the manufacturing strategy process. Part III is a reference that can be consulted as required during the formulation of the manufacturing strategy.

Part I Elements in Manufacturing Strategy	**Part II** Changes and Techniques Frequently Used in Manufacturing Strategy	**Part III** The Seven Production Systems
1. Manufacturing outputs	10. Focus 11. Learning, the product life cycle	
2. Production systems	10. Focus 11. Learning, the product life cycle 12. Evaluation of investments in manufacturing	13. Job shop 14. Batch flow 15. Flexible manufacturing system 16. Operator-paced line flow 17. Just-in-time 18. Equipment-paced line flow 19. Continuous flow
3. Manufacturing levers	9. Improvement approaches 10. Soft and hard technologies 12. Evaluation of investments in manufacturing	
4. Competitive analysis	8. Integrating manufacturing strategy with business strategy 9. Benchmarking 11. The product life cycle	
5. Level of manufacturing capability	9. Benchmarking, improvement approaches 11. Learning	
6. Complete framework for formulating manufacturing strategy		
7. Complete framework for developing the implementation plan	9. Benchmarking, improvement approaches	
8. Integrating manufacturing strategy with business strategy		

Note: Numbers refer to chapter numbers.

FIGURE C

How This Book Is Organized

Further Reading

Adesso, G. A., "Competitive Manufacturing in the Eighties." *Strategic and Tactical Issues in Just-in-Time Manufacturing: Proceedings of the 1985 Annual Conference of the Association for Manufacturing Excellence*, J. B. Dilworth (ed.), pp. 65–76, 1985.

Baumann, L. J., "IBM Manufacturing Strategies for Competitiveness," presented at the *Second Annual Manufacturing Forum*, University of Michigan, Ann Arbor, MI, November 22, 1988.

Beckman, S. L., W. A. Boller, S. A. Hamilton, and J. W. Monroe, "Using Manufacturing as a Competitive Weapon: The Development of a Manufacturing Strategy," pp. 53–75 in *Strategic Manufacturing*, P. E. Moody (ed.), Homewood, IL: Dow-Jones Irwin, 1990.

Fine, C., and A. Hax, "Manufacturing Strategy: A Methodology and an Illustration," *Interfaces*, Vol. 15, No. 6, pp. 15–27, 1985.

Hayes, R. H., and S. C. Wheelwright, *Restoring Our Competitive Edge: Competing Through Manufacturing*, New York: John Wiley & Sons, 1984.

Hayes, R. H., S. C. Wheelwright, and K. B. Clark, *Dynamic Manufacturing: Creating the Learning Organization*, New York: The Free Press, 1988.

Hill, T., *Manufacturing Strategy: Text and Cases*, Homewood, IL: Richard D. Irwin, 1989.

Leong, G. K., D. L. Snyder, and P. T. Ward, "Research in the Process and Content of Manufacturing Strategy," *Omega: The International Journal of Management Science*, Vol. 18, No. 2, pp. 109–122, 1990.

Moody, P. E., "Digital Equipment Corporation: Journeying to Manufacturing Excellence," pp. 175–186 in *Strategic Manufacturing*, P. E. Moody (ed.), Homewood, IL: Dow-Jones Irwin, 1990.

PART I

ELEMENTS IN MANUFACTURING STRATEGY

CHAPTER 1

MANUFACTURING OUTPUTS

Provide better products in a wider variety and at a lower cost. And do it faster. This is the directive given to manufacturing managers all over the world. Customers are demanding more. Competitors are providing more. Manufacturing must do more and do it with less. But what does manufacturing provide to its customers? Manufacturing provides six manufacturing outputs—cost, quality, performance, delivery, flexibility, and innovativeness—to its

SITUATION 1.1

Water Heater Producer Faces Competition

TOP MANAGEMENT of a North American producer and distributor of commercial and residential water heaters worried that manufacturing could no longer provide what was needed to satisfy the organization's customers. Manufacturing provided what it had always provided—a traditional range of products at a cost that had not increased in many years; with quality levels that were as high as ever, at its normal lead time, and with its usual conditions for scheduling, volume changes, and product design changes. However, what was acceptable in the past was not good enough today. A new competitor had entered part of the market and was taking away market share by providing products with lower cost and faster delivery.

customers. Some outputs will be provided at higher levels than others because no single production system can provide all outputs at the highest possible levels. Consequently a plan or strategy is needed to determine precisely how the required outputs will be provided at the required levels.

SITUATION 1.2

Chip Maker Considers Manufacturing Improvement Options

VARIOUS STAFF members and line groups in a company that designed, manufactured, and distributed integrated circuit chips were submitting proposals for improving manufacturing operations. These included proposals to implement new manufacturing technologies and improvement approaches such as setup time reduction, statistical quality control, just-in-time, kaizen, and MRP II. Top management was bewildered by the many technologies and changes. Which were most appropriate for the company? In what order should they be implemented? How quickly should each be implemented? What would the benefits be?

THE SIX MANUFACTURING OUTPUTS

Ask most people what manufacturing provides to its customers and they respond, low-cost, high-quality products. While this view was suitable in the past, it is too simplistic and restrictive for most organizations today. Manufacturing provides six outputs—cost, quality, performance, delivery, flexibility, and innovativeness. These six manufacturing outputs are defined in Figure 1-1. (The appendix at the end of this chapter gives a historical perspective on the development of the six manufacturing outputs.)

COST

Each product that manufacturing produces has a cost. All things being equal, a low cost gives a low price and provides a better opportunity for profit than does a high cost. Product cost is a straightforward concept, but it can be tricky to measure, especially when manufacturing has large overhead costs that need to be allocated. The more important cost is, relative to the other manufacturing outputs, the greater will be the investment in a manufacturing cost accounting system to track and measure it.

Cost	The cost of material, labor, overhead, and other resources used to produce a product.
Quality	The extent to which materials and operations *conform to specifications and customer expectations,* and how tight or difficult the specifications and expectations are.
Performance	The *product's features,* and the extent to which the features or design permit the product to do things that other products cannot do.
Delivery time and delivery time reliability	The time between order taking and delivery to the customer. How often are orders late, and how late are they when they are late?
Flexibility	The extent to which volumes of *existing products* can be increased or decreased to respond quickly to the needs of customers.
Innovativeness	The ability to quickly introduce *new products* or make design changes to existing products.

FIGURE 1–1

Manufacturing Outputs: What Manufacturing Provides Its Customers

QUALITY AND PERFORMANCE

Many nonmanufacturing people view quality and performance as a single manufacturing output. For example, a Mercedes-Benz automobile is said to have outstanding quality. When this means that a Mercedes-Benz is designed and manufactured to exacting specifications, this satisfies the definition of the quality. When it means that a Mercedes-Benz automobile has unique design features, such as a heavier frame, thicker upholstery, a more durable engine and transmission, a better radio, and a richer coat of paint, then we have the performance output.

In manufacturing strategy, quality is associated with conformance to specifications and critical customer expectations. Performance is associated with features of the product as they affect the product's ability to do what other products cannot. While separating quality and performance into two manufacturing outputs may seem picky, the distinction is very important for good manufacturers. Tools and technologies that provide high levels of quality (such as statistical quality control and standardization) are often different from those that provide high levels of performance (such as concurrent engineering and highly skilled workers). Differentiating between quality and performance allows us to design one production system for a McDonald's restaurant (where quality is important) and a different production system for a fine French restaurant (where performance is important).

Delivery

The delivery manufacturing output comprises delivery time and delivery time reliability. Delivery time is the amount of time a manufacturer requires to supply a product to a customer. Usually delivery times are well known and are used to make delivery promises to customers when they place their orders. Often, especially in busy times, manufacturing cannot meet delivery dates and customers are told that they will receive their orders later than promised. When this happens, delivery time reliability drops. Customer expectations for delivery time and delivery time reliability have increased dramatically in recent years as a consequence of the development of just-in-time. Customers now expect to be supplied on time and in small lots.

Flexibility and Innovativeness

Like quality and performance, flexibility and innovativeness are often treated as a single manufacturing output by those outside manufacturing. However, good manufacturers must differentiate between the two. To illustrate the difference, consider the apparel industry. A clothing manufacturer is flexible when he/she can easily change product mix and production volumes in response to changes in fashion and season. This is flexibility because it is the ability to increase or decrease production of *existing products*. Innovativeness is the ability to produce *new products*. A tailor is more innovative than flexible. A tailor has little difficulty producing a suit from a new design. It is more difficult, however, to double the number of suits produced in a month. Differentiating between flexibility and innovativeness helps good manufacturers design and manage production systems that will provide high levels of whichever output is most important to their customers.

Trade-Offs

Much to the surprise of those in manufacturing and to the disappointment of those in marketing, no manufacturer in the world is able to provide all six outputs at the highest possible levels. For example, it may cost a little more to have flexible workers. Standardizing product design may be necessary to achieve the lowest possible cost, even though it reduces innovativeness. Delivery may be longer if products with the newest features (that is, performance) are required.

Further evidence that trade-offs must be made is given in Figure 1-2. Because they cannot provide all six outputs at the highest possible levels, these well-known manufacturers have chosen, either implicitly or explicitly, to provide a single output at the highest possible level in the world while still providing the other outputs at relatively high levels. The products produced by

FIGURE 1–2

Products Well Known for a Particular Manufacturing Output

Manufacturing Output Providing a Competitive Advantage	Product	Advertising Slogan
Cost	BIC pens and razors Commodities and generic products Steel bars and shapes	
Quality	Ford McDonald's restaurants Toyota	"Quality is job one." "There's quality and there's Toyota quality."
Performance	Acura Legend car Christian Dior Chrysler Imperial car Crest toothpaste Honda power equipment Makita power tools Mercedes-Benz Sailing yachts Tide detergent Toyota	"Precision crafted performance." "A name synonymous with quality." "There's no luxury without engineering." "Crest reduces cavities." "Honda engineering makes the difference." "More torque to do the job." "Engineered like no other car in the world." "If it's got to be clean, it's got to be Tide." "I love what you do for me."
Delivery	Airlines Federal Express Photo-finishing	"Now Federal Express gives you a new way to ship overnight; introducing afternoon pickup."
Flexibility	Clothing manufacturers DAF trucks Furniture Housing Industrial machinery Machine shops Tool and die shops	"1987 European Transport Truck of the Year."
Innovativeness	3M DuPont Hewlett-Packard Sony audio/video products BASF	"3M – Making innovation work for you." "Better things for better living." "The spirit of innovation. We don't make the products you buy. We make the products you buy better."

these manufacturers are known in the marketplace for the high level at which a particular manufacturing output is provided. This competitive advantage is actively promoted in advertisements and other marketing activities. The production systems used to manufacture these products are carefully designed and managed so that the desired outputs are provided at the highest possible levels.

Competing on Cost

Many commodities such as structural steel shapes, chemicals, electronic devices, paper, and so on, are manufactured to standard specifications and sold under standard delivery terms. For these products, a customer's purchase decision is based exclusively on price. Consequently, cost becomes the most important manufacturing output.

Competing on Quality

One of the major reasons for the success of McDonald's restaurants is that customers can rely on receiving identical products in every restaurant. Each product in the limited product line is produced to the same exacting specifications. Quality is provided at the highest possible level. Cost and delivery are also important but always give way to quality.

Competing on Delivery

As the following two examples show, many manufacturers like to compete on the basis of delivery. A Westinghouse division that produced custom-engineered electrical distribution equipment decided to make delivery the most important manufacturing output for an important product family. Products were redesigned to be produced from modular components (somewhat reducing innovativeness). The components were stocked in inventory (somewhat increasing cost). When orders were received, the modular components were assembled quickly into final products and shipped to customers (significantly improving delivery time).

A North American manufacturer of cellular telephones with manufacturing facilities in an offshore, low-wage country was having trouble meeting its customers' requirements for fast, reliable delivery. The customers were large retailers who wanted fast delivery of medium quantities of cellular phones with special features. Because these manufacturing outputs could not be provided from an offshore facility, the company decided to move its manufacturing operations to a new facility in North America, that was designed to provide the highest possible levels of delivery and flexibility.

Competing on Performance

When discussing manufacturing outputs with a group of manufacturing executives, one Procter & Gamble executive stated without hesitation that Tide detergent competed on the basis of performance. "Tide got clothes cleaner." It was designed with features that permitted it to do something other products could not.

Evans and Lindsay (1992, p. 149) describe three kinds of product features: dissatisfiers, satisfiers, and exciters/delighters. Dissatisfiers are features that customers expect in a product. If they are not present, the customer is dissatisfied. Satisfiers are features that customers want. When they are present, customers are satisfied. Exciters/delighters are new features that customers do not expect. The presence of these features leads to a high regard for the product and a high level of satisfaction. All production systems must provide the dissatisfiers and the satisfiers. Only those capable of providing a high level of performance will provide the exciters/delighters. Note that it is not difficult to provide a high level of performance occasionally. What is difficult is providing a high level of performance year in and year out. One reason is that exciters/delighters become satisfiers (and satisfiers become dissatisfiers), so new exciters/delighters must be developed continually.

Competing on Flexibility

On a tour of the DAF engine plant and assembly plant in Europe, a DAF materials manager took two large brackets from different bins and asked whether the group could detect any difference between them. Except for slight differences in the bracing and in the location of a few bolt holes, the parts were identical. The materials manager went on to say that this was an example of why more standardization of parts and components was necessary at DAF. Standardization would reduce the large number of different parts that needed to be stocked and controlled. It would also reduce the floor space needed for inventory. All of this would reduce costs.

Before commenting on the importance of standardization compared to other improvements that could be made, the group decided to complete the tour. By the end of the tour, it was obvious that customers bought DAF transport trucks because the company provided an almost unlimited number of options. This allowed customers from all over Europe to obtain the truck that was perfect for their requirements. And DAF did this well. In 1987 it won the prestigious "European Transport Truck of the Year" award. For manufacturing to provide such a high level flexibility, it may be necessary for DAF to have many more part numbers than its competitors. The tour group concluded that, while it is always important to guard against a proliferation in the number of part numbers, it was even more important to remember that DAF's competitive advantage was flexibility. DAF was really too small to compete on the basis of cost in the standard transport truck business, where two giants, Mercedes and Volvo, dominated the market.

Competing on Innovativeness

Sony and 3M are probably the best examples of companies that compete by providing innovativeness at the highest possible level. Innovativeness is the ability to effectively introduce new products and make design changes to existing products. Sony and 3M always seem to be first in their industries to introduce new products. They make trade-offs that favor innovativeness—making it easier to introduce new products. One example of this was Sony's introduction of the first home VCR in the early 1980s. At that time, the VCR was considered a complex electromechanical product. Sony's design staff and manufacturing staff decided to design and produce a product that was considerably larger than necessary from a design point of view. Although cost would increase, the larger size would make the product easier to manufacture, thus permitting the product to be brought to the market sooner, and giving Sony a chance to gain a large share of the market before competitors introduced their products. It would also increase the quality and reliability (a measure of performance) of the new product, which were necessary to ensure quick customer acceptance.

Meeting and Exceeding Customer Expectations

The manufacturing outputs required by customers and the levels at which they are required change over time. Manufacturing cannot provide all six outputs at the highest possible levels. Hence, it is important for manufacturing to determine which outputs are most important to customers now, and which will be important in the future. This is called meeting and exceeding customer expectations. For example, when Japanese companies entered the North American color television market in the 1970s, they changed the basis of competition from price to quality. High levels of quality attributes, such as picture clarity, time without breakdowns, and so on, were required to win orders. North American manufacturers struggled to improve their production systems so that this high level of quality could be provided. As soon as this was accomplished, however, the basis of competition changed from quality back to price.

Meeting and exceeding customer expectations for the purpose of gaining a competitive advantage is the goal of manufacturing strategy. The goal is achieved when manufacturing provides the required manufacturing outputs at the required levels. Before setting out to do this, the organization should realize that manufacturing cannot provide all six outputs at the highest possible levels. Trade-offs must be made. A manufacturing strategy permits systematic trade-offs. It takes into account the requirements of customers as well as what is provided by competitors. It considers manufacturing's current capabilities and the full range of options available to manufacturing. When it is finished, the strategy specifies the levels at which each manufacturing output will be provided and how manufacturing will be changed to accomplish this.

Meeting and exceeding customer expectations is also how quality is defined in the total quality management (TQM) improvement approach. TQM takes the six outputs—cost, quality, performance, delivery, flexibility, and innovativeness—that manufacturing provides for its customers and groups them into a single output called quality, the level of which must be sufficiently high to meet and exceed customer expectations. While this grouping is satisfactory for TQM (see Chapter 9), it is too coarse to be of practical use in manufacturing strategy. For one thing, it does not facilitate the making of precise decisions and trade-offs.

SUMMARY

Manufacturing provides six manufacturing outputs to its customers: cost, quality, performance, delivery, flexibility, and innovativeness. For each product or product family, one or two outputs are provided at the highest possible level in the world, while the others are provided at high, but somewhat lower, levels. This gives a basis of competitive advantage while recognizing that no single production system can provide all outputs at the highest possible levels. A plan or strategy is needed for manufacturing to determine precisely how the required outputs will be provided at the required levels. The elements of this manufacturing strategy are presented in the next four chapters.

FURTHER READING

Evans, J., and W. Lindsay, *The Management and Control of Quality*, Second Edition, Minneapolis: West Publishing, 1992.

Fine, C., and A. Hax, "Manufacturing Strategy: A Methodology and an Illustration," *Interfaces*, Vol. 15, No. 6, pp. 15–27, 1985.

Hayes, R., and S. Wheelwright, *Restoring Our Competitive Edge: Competing Through Manufacturing*, New York: John Wiley & Sons, 1984.

Hill, T., *Manufacturing Strategy: Text and Cases*, Homewood, IL: Richard D. Irwin, 1989.

Wheelwright, S., "Manufacturing Strategy: Defining the Missing Link," *Strategic Management Journal*, Vol. 5, pp. 77–91, 1984.

APPENDIX: A HISTORICAL PERSPECTIVE ON THE SIX MANUFACTURING OUTPUTS

In the 1960s and early 1970s, most people felt that manufacturing provided only two outputs to their customers—cost and quality; a notable exception was Wickham Skinner (1969). During the 1980s the list of manufacturing outputs increased with the addition of delivery and flexibility (Fine and Hax 1985; Stalk 1988; Hall and Nakane 1990). This was mostly a consequence of the development of just-in-time techniques, the just-in-time production system, and flexible manufacturing systems. Quickly, flexibility became a catch-all output. Included in its definition were (Browne et al. 1984):

- *Machine flexibility*: The ease of changing tools, adjusting machine settings, making repairs, changing NC programs, and so on.
- *Process flexibility*: The ability to produce products in different ways.
- *Product flexibility*: The ability to change over to produce a different product.
- *Routing flexibility*: The ability to produce products on different machines in the event of a breakdown.
- *Volume flexibility*: The ability to operate machines profitably at different volumes.
- *Expansion flexibility*: The ability to expand a system of machines to create more capacity.
- *Operation flexibility*: The ability to change the order in which operations are done for each product.
- *Production flexibility*: The universe of all products that the machines can produce.

The definition was so broad that the usefulness of flexibility as an important consideration in manufacturing strategy was greatly diminished.

Flexibility has been redefined. Those types of flexibility associated with existing products are still called flexibility, but those types of flexibility associated with the introduction of new products and design changes to existing products comprise a new manufacturing output called innovativeness.

Similar developments occurred with quality. Definitions such as "quality is giving the customer what he or she wants" and "quality is meeting and exceeding customer expectations" are too broad to be useful for manufacturing strategy. It is more useful to think of two manufacturing outputs: quality and performance. Quality is associated with conformance to specifications and meeting critical customer expectations. Performance is associated with features of the product as they affect the product's ability to do what other products cannot.

There is considerable empirical evidence that managers view the original four manufacturing outputs—cost, quality, delivery, and flexibility—as very important in manufacturing strategy. (See Schroeder, Anderson, and Cleveland 1986; Huete and Roth 1987; and Roth and Miller 1990.) This is not yet the case for innovativeness and performance. There are three reasons for this: 1) Many organizations are

FIGURE 1-3

Operational Measures of Manufacturing Outputs

Dimension	Original Variable	Equivalent Manufacturing Output in This Book
Price	Offer low prices	Cost
Quality (.70)[1]	Consistent quality; reliability (.79)[2] High-performance products (.76)	Quality Performance
Delivery (.77)	Fast deliveries (.83) Dependable delivery promises (.88)	Delivery time Delivery time reliability
Flexibility (.53)	Rapid volume changes (.52) Design changes; introduce new products (.92)	Flexibility Innovativeness
Market scope (.74)	Broad product line (.64) Broad distribution (.86) Advertise effectively (.78) After-sales service (.62)	Flexibility Nonmanufacturing outputs provided by other functional areas (see Chapter 8)

Notes

1. Cronbach alpha, a measure of the reliability of the dimension.
2. The correlation coefficient of the original variable and the dimension.

– The higher the Cronbach alpha and the correlation coefficient, the stronger is the relationship between the original variable and the dimension. A value over .6 is excellent.

Source: 1988 data from a study by Roth and Miller 1990.

struggling to provide high levels of cost, quality, delivery, and flexibility. Only after they achieve this will they switch their attention to innovativeness and performance. 2) In many organizations, innovativeness and performance are included in the flexibility and quality outputs. 3) It was not until the 1990s that innovativeness and performance emerged as important manufacturing outputs. This development would not be reflected in the most recent empirical evidence, which comes from the late 1980s.

Consider the results in Figure 1-3, taken from a study by Roth and Miller (1990). In this part of the study, eleven variables measuring manufacturing success are grouped into five aggregate variables called dimensions. The numbers in parentheses (Cronbach alphas and correlations) suggest that this is an effective grouping. If we examine the eleven variables, we notice that they could also be grouped into the six manufacturing outputs used in this book. Exactly how efficient this grouping would be (that is, what the Cronbach alphas and correlations would be) is a subject for future analysis with more recent data.

Many ways of measuring the six manufacturing outputs have been developed. Leong, Snyder, and Ward (1990) compiled a list of measures, which is shown with some newer measures in Figure 1-4. These measures will be useful in the competitive analysis element of manufacturing strategy (Chapter 4).

FIGURE 1–4

Measures for the Six Manufacturing Outputs

Manufacturing Output	Measures	Source
Cost	Unit product cost	Fine and Hax (1985); Roth (1989)
	Unit labor cost	Fine and Hax (1985); Roth (1989)
	Unit material cost	Fine and Hax (1985); Roth (1989)
	Total manufacturing overhead cost	Roth (1989)
	Inventory turnover — raw material, WIP, finished goods	Fine and Hax (1985); Roth (1989)
	Capital productivity	Fine and Hax (1985)
	Capacity/machine utilization	Roth (1989)
	Materials yield	Roth (1989)
	Direct labor productivity	Roth (1989)
	Indirect labor productivity	Roth (1989)
Quality	Internal failure cost — scrap and rework, percentage defective or reworked	Fine and Hax (1985); Garvin (1986)
	External failure cost — frequency of failure in the field	Fine and Hax (1985); Garvin (1986)
	Quality of incoming material from suppliers	Roth (1989)
	Warranty cost as a percentage of sales	
	Rework cost as a percentage of sales	
Performance	Number of standard features	
	Number of advanced features	
	Product resale price	
	Number of engineering changes	Roth (1989)
	Mean time between failures	Fine and Hax (1985)
Delivery	Quoted delivery time	De Meyer et al. (1989)
	Percentage of on-time deliveries	Fine and Hax (1985); Roth (1989)
	Average lateness	Fine and Hax (1985)
	Inventory accuracy	Roth (1989)
	Master production schedule performance/stability	Roth (1989)
Flexibility	Number of products in the product line	
	Number of available options	
	Minimum order size	
	Average production lot size	
	Length of frozen schedule	
	Number of job classifications in the factory	
	Average volume fluctuations that occur over a time period divided by the capacity limit	Gerwin (1987)
	Number of parts processed by a group of machines	Gerwin (1987)
	Ratio of number of parts processed by a group of machines to the total number processed by the factory	Gerwin (1987)
	Number of setups	Gerwin (1987)
	Variations in key dimensional and metallurgical properties that can be handled by the equipment	Gerwin (1987)
	Is it possible to produce parts on different machines	Gerwin (1987)
Innovativeness	Number of engineering change orders per year	
	Number of new products introduced each year	
	Lead time to design new products	
	Level of R&D investment	Mansfield (1981)
	Consistency of investment over time	Maidique and Hayes (1984)

Source: Adapted in part from Leong, Snyder, and Ward 1990.

References

Browne, J., K. Rathmill, S. Sethi, and K. Steche, "Classification of Flexible Manufacturing Systems," *The FMS Magazine*, pp. 114–117, April 1984.

De Meyer, A., J. Nakane, J. Miller, and K. Ferdows, "Flexibility: The Next Competitive Battlefield," *Strategic Management Journal*, Vol. 10, pp. 135–144, 1989.

Fine, C., and A. Hax, "Manufacturing Strategy: A Methodology and an Illustration," *Interfaces*, Vol. 15, No. 6, pp. 28–46, 1985.

Garvin, D. A., "Quality Problems, Policies and Attitudes in the United States and Japan: An Exploratory Study," *Academic Management Journal*, Vol. 29, No. 4, pp. 653–673, 1986.

Gerwin, D., "An Agenda for Research on the Flexibility of Manufacturing Processes," *International Journal of Operations and Production Management*, Vol. 7, No. 1, pp. 38–49, 1987.

Hall, R. W., and J. Nakane, *Flexibility: Manufacturing Battlefield of the 90s. A Report on Attaining Manufacturing Flexibility in Japan and the United States*, Wheeling, IL: Association for Manufacturing Excellence, 1990.

Huete, L., and A. V. Roth, "Linking Manufacturing Capabilities with SBU Strategic Decisions," *Proceedings of the 1987 Decision Sciences Conference*, 1987.

Leong, G. K., D. L. Snyder, and P. T. Ward, "Research in the Process and Content of Manufacturing Strategy," *Omega: The International Journal of Management Science*, Vol. 18, No. 2, pp. 109–122, 1990.

Maidique, M. A., and R. H. Hayes, "The Art of High Technology Management," *Sloan Management Review*, Vol. 25, No. 2, pp. 17–31, 1984.

Mansfield, E., "Composition of R&D Expenditure: Relationship to Size of Firm, Concentration and Innovative Output," *Rev. Economics Statistics*, Vol. 63, No. 4, pp. 610–615, 1981.

Roth, A. V., "Linking Manufacturing Strategy and Performance: An Empirical Investigation," Working Paper, Boston University, 1989.

Roth, A. V., and J. G. Miller, "Manufacturing Strategy, Manufacturing Strength, Managerial Success, and Economic Outcomes," *Manufacturing Strategy*, J. E. Ettlie, M. C. Burstein, A. Feigenbaum (eds.), Dordrecht, The Netherlands: Kluwer Publishers, pp. 97–108, 1990.

Schroeder, R. G., J. C. Anderson, and G. Cleveland, "The Content of Manufacturing Strategy: An Empirical Study," *Journal of Operations Management*, Volume 6, pp. 405–415, August 1986.

Skinner, W., "Manufacturing—Missing Link in Corporate Strategy," *Harvard Business Review*, pp. 136–145, May–June 1969.

Stalk, G., "Time—The Next Source of Competitive Advantage," *Harvard Business Review*, pp. 41–51, July–August 1988.

CHAPTER 2

PRODUCTION SYSTEMS

How does manufacturing provide the cost, quality, performance, delivery, flexibility, and innovativeness outputs to its customers? The answer is obvious. We see it each time we drive down the highway and pass factories and offices with employees streaming in and out, suppliers making deliveries, and shippers moving finished goods. We know that inside these buildings are machines and processes, workers and managers, departments and control systems: everything working together to form a production system. The production system provides the cost, quality, performance, delivery, flexibility, and innovativeness manufacturing outputs.

There are both similarities and differences between the production systems used at different companies. Two factors determine the extent of the similarities and differences; the type of product manufactured, and the manufacturing outputs provided. For example, the production systems used at ISTC and Dofasco are extremely different. Each company manufactures a different type of product and provides different manufacturing outputs to its customers. ISTC manufactures railroad and subway cars. Dofasco, a fully integrated steel producer, is one of North America's most profitable steel companies. Because ISTC's products are highly engineered and custom designed for specific customer orders, the ISTC production system must provide high levels of innovativeness and performance. The production systems used by other manufacturers who also provide high levels of innovativeness and performance (such as manufacturers of machine tools, industrial packaging equipment, and machine shops supplying parts in the aerospace industry) are very similar to ISTC's production system. Dofasco, on the other hand, produces large volumes of standard products on a production system that provides high levels of quality, cost and delivery. Paper mills, container manufacturers, refineries, and brewers all have production systems like the one at Dofasco.

Thus, it is not surprising that there are only a few different production systems—seven in all. This has a major implication for manufacturing strategy.

> The production system used by an organization should be the one that is best able to provide the manufacturing outputs demanded by the organization's customers.

When this is not the case, the production system should be changed to a more appropriate one, or, at the very least, the manufacturer should realize that the company is vulnerable to any competitor who has a more appropriate production system. We will return to this in Chapter 4. For now, we will examine the seven different production systems.

FIGURE 2–1

The Seven Production Systems

Production System	Product / Volume	Layout / Flow
Traditional systems		
Job shop	Very many products / One or a few of each	Functional layout / Flow extremely varied
Batch flow	Many products / Low to medium volumes	Cellular layout / Flow varied with patterns
Operator-paced line flow	Several to many products / Medium volumes	Line layout / Flow mostly regular, paced by operators
Equipment-paced line flow	Several products / High volumes	Line layout / Flow regular, paced by equipment
Continuous flow	One or a few products / Very high volumes	Line layout / Flow rigid, continuous
New Systems		
Just-in-time (JIT)	Many products / Low to medium volumes	Line layout / Flow mostly regular, paced by operators
Flexible manufacturing system (FMS)	Very many products / Low volumes	Cellular or line layout / Flow mostly regular, paced by equipment

The Seven Production Systems

There are only seven different production systems—five traditional systems and two new systems. They are listed in Figure 2-1. The product/volume–layout/flow (PV–LF) matrix shown in Figure 2-2 is a useful tool for analyzing the relationships among the seven production systems. The PV–LF matrix has four dimensions:

1. The number of *products* produced
2. The production *volume* of each product
3. The *layout* or arrangement of equipment and processes used to manufacture the products
4. The *flow* of material through the equipment and processes

The PV–LF matrix builds on the product-process matrix developed by Robert Hayes and Steven Wheelwright in 1979.

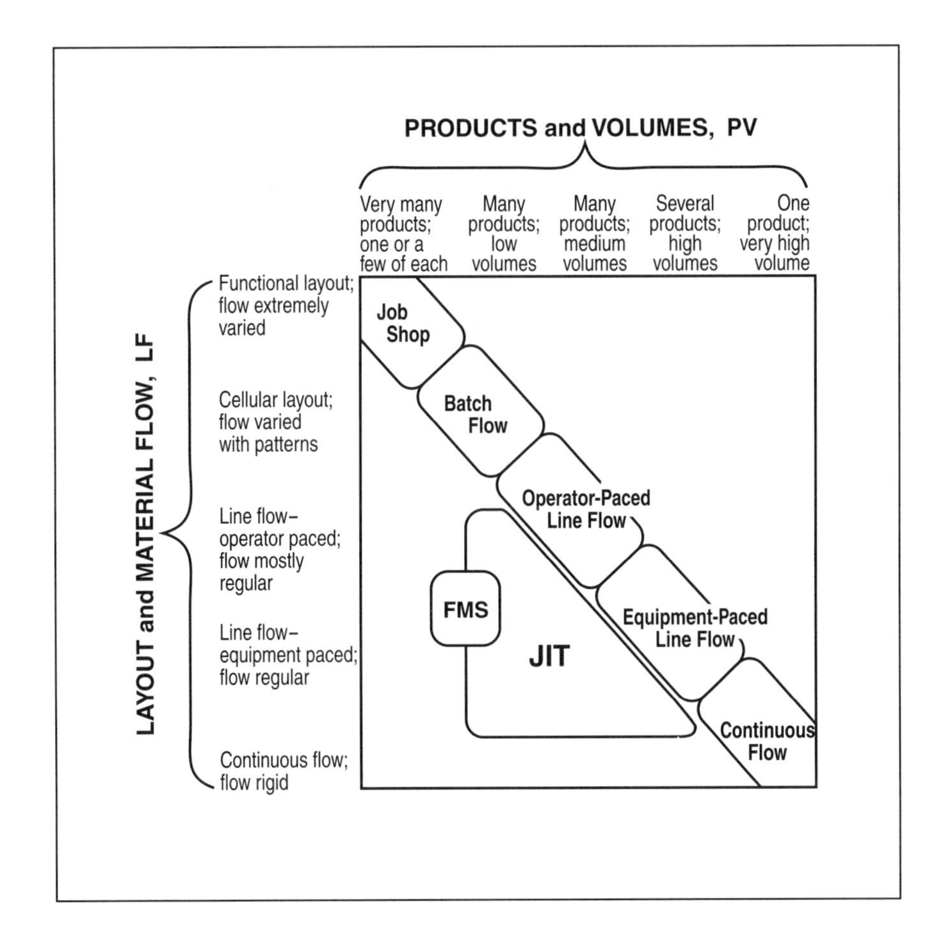

FIGURE 2–2

The Product/Volume–Layout/Flow (PV–LF) Matrix

Hayes and Wheelwright proposed that many characteristics of production units varied with two primary dimensions—product structure and process structure. They also proposed that these two dimensions were related to each other as a consequence of shared life cycles. (See Chapter 11.) Finally, they outlined a relationship between the dimensions and some simple characteristics of production units. Their product-process matrix has proven to be useful and is now widely used. (See the readings at the end of the chapter.)

Products/Volumes

The first two dimensions on the PV–LF matrix are shown across the top of the matrix (Figure 2-2). A measure for these dimensions is the following product/volume question:

> What products are manufactured, and in what volumes are they manufactured?

The answer ranges from, "We produce whatever products our customers ask for, in whatever volume they want" to "We produce a very high volume of one product." These two answers anchor the range of possible answers to the product/volume question. Other points along the range are:

- Very many different products produced in volumes of one, or a few, of each product
- Many different products produced in low volumes
- Many different products produced in medium volumes
- Several different products produced in high volumes
- One product produced in very high volume

Manufacturers have little difficulty answering the product/volume question. One Tridon Ltd. plant produced a family of hose clamps and a family of automotive electrical switches. The hose clamp family consisted of a few products produced in very high volumes (tens of millions per year), while the electrical switch family consisted of several different products produced in low volumes. A visit to the plant showed that there were two "plants-within-the-plant" (Chapter 10), one for clamps and one for switches, and that each plant-within-the-plant used a different production system.

LAYOUT/FLOW

The other two dimensions are tracked down the side of the PV–LF matrix and are measured by asking the layout/flow question:

> How are the equipment, processes, and departments arranged in the plant, and how does material flow from workstation to workstation as it moves through the plant?

Plant Layout

There are three basic plant layouts (Figure 2-3):

- Functional layout
- Cellular layout
- Line layout

The layouts used in a particular plant are easily gleaned from an examination of a drawing of the plant layout.

In a *functional layout*, equipment of the same type is located in the same area. For example, in a machine shop, lathes may be located in one department, shears in another, welders in yet another department, and so on. Operators and supervisors work in one department and are highly skilled on the type of equipment in the department. Equipment and tooling are general purpose and are capable of performing a wide range of operations.

In a *cellular layout*, different types of equipment and processes are located in the same department so that all the operations required to produce any product within a relatively large product family can be done in the department. (Departments are sometimes called manufacturing cells.) Operators are trained to operate all the equipment in the department. Equipment and tooling are general purpose because they are used to produce all the products in the large product family, but it is often possible to specialize them a little.

In a *line layout*, the different types of equipment needed to produce one product or a small product family are arranged into a line. Equipment and tooling are designed for the product or product family being produced. Line layouts are used when the volume is sufficiently high to justify the use of expensive, dedicated equipment. Because equipment does most of the work, operators are left to perform relatively simple tasks.

FIGURE 2–3

The Three Basic Layouts

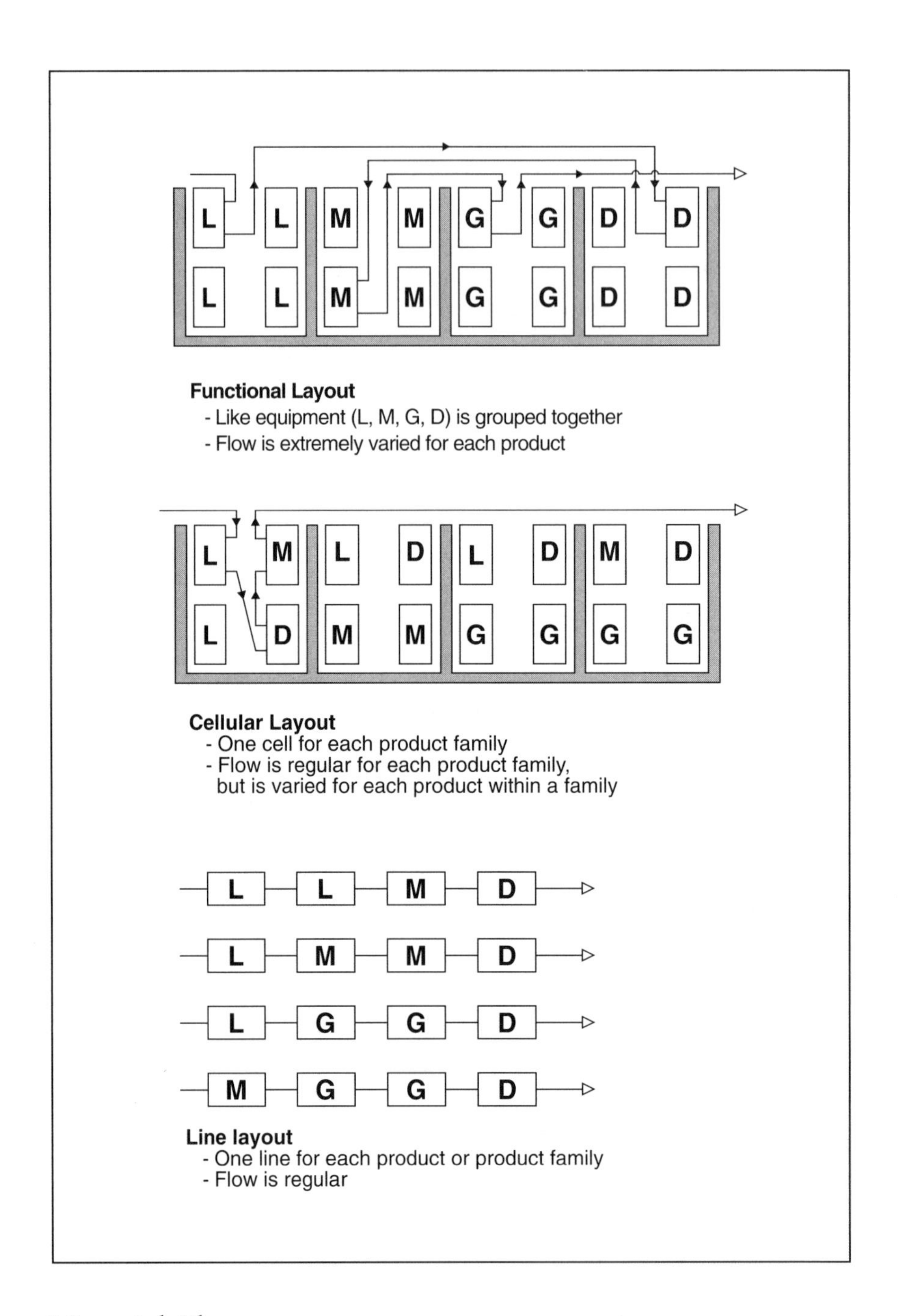

Material Flow

To determine the material flow, walk through the plant, starting at the unloading docks where purchased material and parts are received, and continue through the plant along the path taken by material as it moves from workstation to workstation. The walk ends at the work center where the finished product is packaged and loaded for shipment to customers. The material flow depends on the layout, but for a particular layout, it can still vary somewhat. The range of plant layouts and material flows is shown in Figure 2-2.

When a plant is organized into a functional layout, the material flow is extremely varied. Depending on what operations need to be performed, the order in which they need to be done, and the availability of equipment and operators, the material flow can vary considerably from order to order. In a plant with a cellular layout, the flow is more regular because each product or product family is always processed in a particular department or cell. Within the cell, however, the material flow can vary from order to order depending on the operations that need to be done, the workload at the machines, and so on.

Many organizations use line layouts; thus it is useful to separate lines into those where the pace or speed of the line is set by the operators and those where the pace is set by the equipment. With operator-paced line flows, the rate at which products are produced depends on the number of operators assigned to the workstations on the line, the speed at which they work, and how well they work together as a team. The material flow in an operator-paced line flow is regular for the most part. There will be some variation when different products are produced, production volumes change, and so on.

SITUATION 2.1

Operator-Paced Line Flow at Lincoln Electric

AN EXAMPLE of an operator-paced line flow is the Lincoln Electric plant in Lincoln, Nebraska, which manufactures welding machines. Materials flow from a long dock on the north side of the plant, through the production lines, to a small storage and loading area on the south side. The production lines are designed so that several different welding machines can be produced on the same line by adjusting the line speed, assigning employees to different workstations, and making use of different tools and fixtures. Per-piece wage rates are paid to each crew, rather than to individual operators, to encourage operators to work together as a team.

An equipment-paced line flow is designed to run at one (or perhaps a few) speed(s) and produce a smaller number of different products than an operator-paced line flow. It is more capital intensive and less flexible, and is designed to run at a higher speed and produce higher volumes. The material flow in an equipment-paced line flow is regular. There may be slight variations when products are produced with different options. Traditional automobile assembly lines are examples of equipment-paced line flows. For example, they may run at a fixed speed of 50 units per hour. There may be slight variations in the material flow when automobiles with different options, such as automatic versus standard transmissions, two-versus four-door models, air-conditioning, and so on, are produced.

Many facilities have material flows that are even more rigid than the equipment-paced line flow. They are called continuous flow lines. Compared to the equipment-paced line flow, the continuous flow line is more capital intensive and more highly automated, requires fewer operators, and allows fewer options to be produced. A continuous flow line can run at only one, extremely high speed. Examples of continuous flow lines are refineries, chemical plants, paper mills, continuous casting and rolling operations in steel mills, and so on. Continuous flow lines produce a very large volume of a single product, and that product is often a commodity.

Each of the seven production systems coincides with a unique set of values for the products, volumes, layout, and flow dimensions (Figures 2-1 and 2-2). More complete descriptions of the production systems are found in Part III of this book. For now, a brief overview of each system will suffice (see Figure 2-4).

Job Shop Production System

The job shop production system produces many different products in volumes ranging from one to a few of each product. The facility has a functional layout, where equipment of the same type is located in the same department. Operators work in one department only and are highly skilled on the equipment in that department. Because many different products are produced in very low volumes, the equipment and tooling are general purpose. The material flow through the job shop varies considerably from job to job, and a great deal of material handling is required to

Production System	Product		Material Flow	Layout	Equipment	Costs		Employees		Organization	
	Variety	Volumes				Fixed	Variable	Staff	Line	Structure	Style
Job shop	Major differences	Very low	Random	Functional	General purpose, flexible	Low	High	Few	Highly skilled	Flat, decentralized	Entrepreneurial
Batch flow	Large variety	Low	Random with patterns	Cells and functional	General purpose, some special-ization	Moderate	Moderate	Few	Multiskilled	Flat, decentralized	Entrepreneurial
Operator-paced line flow	Some variation	Medium	Regular	Line	Specialized, some flexibility	High	Low	Many	Multiskilled	Hierarchy, decentralized	Entrepreneurial
Equipment-paced line flow	Standard, with minor modification	High	Regular	Line	Special purpose	Very high	Low	Many	Unskilled	Hierarchy, centralized	Bureaucratic
Continuous flow	Standard	Very high	Rigid	Line	Special purpose, highly automated	Extremely high	Very low	Very many	Few, unskilled	Hierarchy, centralized	Bureaucratic
Flexible manu-facturing system (FMS)	Major differences	Very low	Regular	Line and cells	Flexible, highly automated	Extremely high	Very low	Many	Few, skilled	Hierarchy, centralized	Bureaucratic
Just-in-time (JIT)	Large variety	Medium to low	Regular	Line	General purpose, many specialized	Moderate	Low	Some	Multiskilled	Flat, decentralized	Entrepreneurial

FIGURE 2–4

Some Characteristics of the Seven Production Systems

move the jobs from department to department. Jobs usually wait a long time before equipment becomes available. Work-in-process inventory is high and delivery times can be long.

Batch Flow Production System

The batch flow production system produces fewer products in higher volumes than the job shop production system. Products are produced in batches, which represent a few months of customer requirements. A combination of functional and cellular layouts is used. Cellular layouts are used when it is cost-effective to place equipment into departments, or cells, to produce families of products. Because there are many products, the equipment and tooling are mostly general purpose. The material flow varies from order to order, although there are patterns of flow for product families and for the larger batches.

Equipment-Paced Line Flow Production System

In line flow production systems, the equipment and processes are arranged into a line and specialized to produce a small number of different products or product families. These systems are used only when the product design is stable and the volume is high enough to make efficient use of a dedicated line.

The equipment-paced line flow production system produces a small number of different products in high volumes at a constant rate. It is capital intensive and extremely specialized. Operators perform relatively simple tasks at a rate determined by the speed of the line. For example, a J-car final assembly line in an automobile plant produces only J-cars (in a limited range of options) at a constant rate of about 60 cars per hour.

Operator-Paced Line Flow Production System

This production system is used when the number of different products is too high and the production volumes are too variable for the equipment-paced line flow production system. The line is more flexible than an equipment-paced line and can be run at a variety of speeds. The rate at which products are produced depends on the particular product being produced, the number of operators assigned to the line, and how well the operators work together as a team.

Continuous Flow Production System

This production system is most similar to the equipment-paced line flow production system. However, it is more automated, more capital intensive, and less flexible. It is designed to produce one product or a narrowly defined product family at very high volumes. The product design is very stable (the product is often a commodity). It competes on the basis of cost and quality, where quality means conformance to standard, industry-wide specifications. This rigid production system, consisting of highly automated, specialized equipment running continuously with little operator assistance, is used to produce the standard product at the lowest possible cost.

Just-in-Time (JIT) Production System

Before describing a JIT production system, it is important to distinguish between JIT techniques and the JIT production system. Many new and not-so-new techniques are called JIT techniques. These include statistical quality control, setup time reduction, multiskilled workers, pull production, standardization, problem-solving, kaizen, and so on. These techniques are used in the JIT production system, but are also used in other production systems. As we will see in Chapter 17, the JIT production system is much more than a collection of JIT techniques. It is a line flow production system that produces many products in low to medium volumes. An operator-paced or an equipment-paced line flow production system cannot be used because there are too many different products and the volumes are too low. By virtue of its design, the JIT production system forces all unnecessary elements (or wastes) to be eliminated. This leads naturally to lower costs, improved quality, and faster delivery.

Flexible Manufacturing System (FMS) Production

The FMS production system consists of computer controlled machines and automatic parts delivery and removal material handling systems, all of which are controlled by a supervisory computer. An FMS can run unattended for long periods. The machines, material handling system, and computers are very flexible, which permits the FMS to produce many different products in low volumes. For example, a Pratt and Whitney FMS, located in Halifax, Nova Scotia, produces 70 different products for the aerospace industry in volumes ranging from 30 to 1,000 units per year. Because an FMS is expensive, it is usually used in situations where simpler, less expensive line flow production systems cannot be used.

Simpler Classifications of Production Systems

It is possible to organize the seven production systems into three groups: craft production, mass production, and lean production (see the book by Womack, Jones, and Roos 1991).

- Craft production
  Job shop production system
  Batch flow production system
- Mass production
  Operator-paced line flow production system
  Equipment-paced line flow production system
  Continuous flow production system
- Lean production
  JIT production system
  FMS production system

The three-group categorization is adequate for many purposes, such as explaining the differences between the new production systems (lean production) and the traditional production systems (craft and mass production). However, the categorization is too broad to be useful for making precise decisions and trade-offs and working out the details of a manufacturing strategy.

Manufacturing Outputs Provided by the Traditional Production Systems

The seven production systems were developed in different parts of the world at different times. There are two important reasons why each of them is still useful today:

1. Each production system provides different levels of the manufacturing outputs—cost, quality, performance, delivery, flexibility, and innovativeness.
2. Each production system is uniquely suited to produce a particular mix and volume of products.

Figure 2-5 shows the levels at which the manufacturing outputs are provided by each production system when the production system is well managed. (The concept of a well-managed production system is discussed in Chapter 5.) The length of the bar is a measure of the level at which an output is provided. Black bars indicate high or good levels; white bars indicate low or poor levels.

Cost and Quality

Production systems near the bottom of the PV–LF matrix can provide better levels of the cost and quality outputs than production systems near the top of the matrix. Production systems such as equipment-paced line flow and continuous flow produce a single product or product family on specialized equipment with specialized tooling using relatively few operators. The specialization ensures that all product specifications are met, no matter how tight they are. Therefore, the quality is high. Because the volume is high, equipment utilization will be high. Consequently, cost is as low as possible.

FIGURE 2–5

Manufacturing Outputs Provided by Well-Managed Production Systems

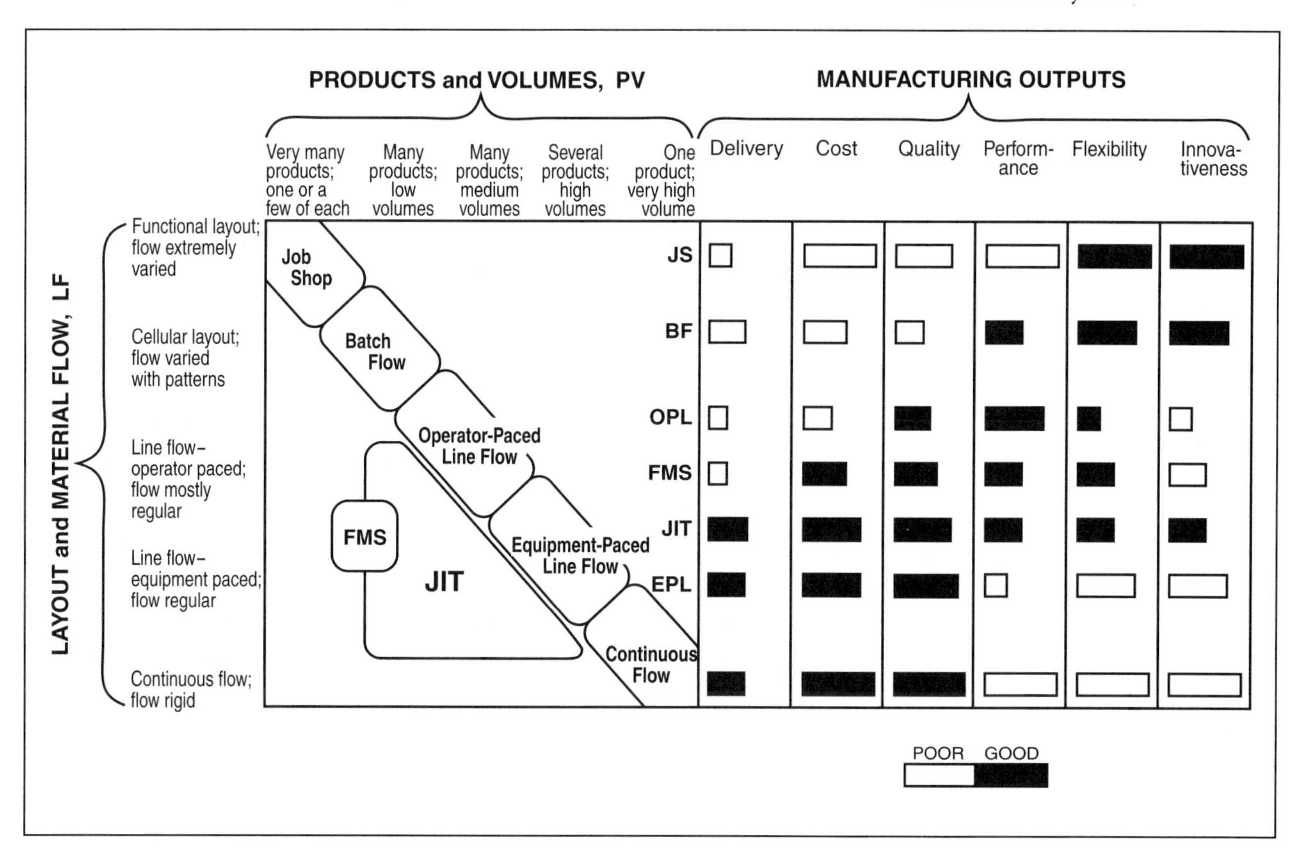

Production systems located near the top of the matrix (job shop and batch flow) produce many different products in relatively small volumes on general-purpose machines with general-purpose tooling. The quality of the products produced by these production systems is good, otherwise customers would not buy them. However, the quality is not as high as it would be if the same products were produced on highly specialized machines with high-volume tooling. The production volumes are too low to permit learning and, hence, improvement to occur. The same reasoning applies to cost. The cost is reasonable (otherwise customers would not buy the products), but it is difficult for a job shop or batch flow production system to match the cost of a line flow production system that produces only a few products in high volumes on specialized equipment.

Note that it is often impossible to produce the same product on production systems from opposite ends of the PV–LF matrix. For example, custom-engineered packaging equipment is produced to customer specifications in a job shop or batch flow production system. Because each order is different and the volumes are low, it is not possible to set up an equipment-paced line flow or continuous flow production system. All companies producing this product will use a job shop or batch flow production system and will thus provide similar levels of the same outputs. Suppose, however, that one competitor sets up an equipment-paced line flow production

SITUATION 2.2

Batch Flow at a Clothing Manufacturer

A CLOTHING manufacturer produces an endless variety of shirts, pants, and jackets on a batch flow production system. Orders, which consist of the type of garment, style, color, and size, are produced in batches of 10 dozen units. The batch flows through a functionally organized plant, with a cutting department, sewing department, pressing department, and so on. The batch is routed from department to department and, within each, to any available workstation until all the required operations are completed. It is relatively easy for this manufacturer to change products and volumes and to introduce new styles because the production system provides high levels of flexibility and innovativeness.

system and produces a family of standard products in high volumes. If the standard product competes directly with any custom-engineered products, then the competitor will have significant cost and quality advantages.

Flexibility and Innovativeness

Production systems near the top of the PV–LF matrix are able to provide better levels of flexibility and innovativeness than production systems near the bottom. Production systems like the job shop and batch flow systems are designed to produce a wide variety of products in relatively low volumes. Machines and tooling are general purpose, and operators are highly skilled. Consequently, it is easy to change volumes, make design changes, and introduce new products.

It is more difficult for an equipment-paced line flow or a continuous flow production system to change products and volumes, make product design changes, and introduce new products. The equipment and tooling are specialized for the product or product family being produced. Design changes and introduction of new products require major adjustments to the equipment and tooling. Consequently, these machine-oriented production systems provide less flexibility and innovativeness than the people-oriented production systems at the top of the PV–LF matrix.

Situation 2.3

Too Much Automation at a Chemical Plant

A COMPANY developed a resin with new advanced features not available in other resins. The company then built a modern continuous flow production system to manufacture the resin. Eighteen months later, these new features became standard features in the industry and were available in the resins produced by most of the company's competitors. At that time, the company could not introduce other new features into its product because of the high cost of converting the rigid, highly automated, continuous flow production system. In hindsight, the company felt that it should have anticipated the need to add new features and built a less rigid, equipment-paced line flow production system.

Delivery

The shortest delivery time and the best delivery time reliability for a particular product are provided by the manufacturer who has a production system dedicated to producing only that product. Equipment-paced line flow and continuous flow production systems, with their specialized equipment and high speeds, will produce a product in the shortest possible time. The delivery time reliability is high because these capital intensive production systems are designed to run for long periods without stopping.

Production systems near the top of the PV–LF matrix can also provide good deliveries for their customers by expediting orders when necessary. However expediting cannot be done all the time. Overall, production systems near the bottom of the PV–LF matrix provide better delivery than the production systems near the top, but the differences are not as great as they are for the cost, quality, flexibility, and innovativeness outputs (Figure 2-5).

Performance

Manufacturers that consistently provide products with a high level of performance have production systems that are not excessively machine-oriented or people-oriented. Well-managed production systems, which rely equally on the capabilities of both machines and people, are best able to provide products with a high level of performance year in and year out.

Situation 2.4

Clothing Manufacturer Changes Production System to Compete on Performance

THE McGREGOR Clothing Company produced socks at a plant located in Toronto—a very expensive city in which to manufacture. Rather than relocate their manufacturing operations to another part of the world and produce standard products on a continuous flow production system, they chose to stay in Toronto, use an operator-paced line flow production system, and compete on the basis of performance. They acquired the rights to produce prominent brands such as Calvin Klein and Christian Dior and produced comfortable clothing made from more expensive materials, designed with generous dimensions and attractive styles.

A high level of performance requires a constant stream of new products, enhancements to existing products, and changes to the process. This is difficult for production systems near the bottom of the PV–LF matrix because they are so rigid. It is costly to change specialized, automated machines and tooling, retrain unskilled operators, stop high-speed lines for a few weeks to make changes, and so on. This can be done from time to time, but it cannot be done at the pace needed to provide a high level of performance year after year. Production systems near the top of the PV–LF matrix have difficulty providing a high level of performance for a different reason. The production volume of any particular product is so low that the organization cannot afford the engineering resources required to design the new, advanced features into the product, or the engineering resources required to design new workstations and processes for producing the new, advanced features.

Manufacturing Outputs Provided by the JIT Production System

A look at Figure 2-5 and the outputs provided by the JIT production system explains why there is so much interest in JIT. The JIT production system is discussed in detail in Chapter 17. For now we simply summarize the reasons why this system can provide high levels of so many outputs.

JIT is a line flow production system and thus can provide the cost, quality, and delivery outputs at the high levels associated with line flow production systems. It does this by exploiting the capabilities of the equipment in the line. At the same time, the JIT production system, like the batch flow system, is designed to produce many products in low to medium volumes. Consequently, flexibility and innovativeness are provided at the high levels associated with the batch flow production system. JIT does this by capitalizing on the unique skills of its workforce.

The JIT production system provides high levels of all manufacturing outputs, but it is the most difficult production system to design and operate. For example, it took Toyota more than 20 years to implement its JIT production system. Although there are many cases of North American and European manufacturers who have successfully implemented several JIT techniques, there are far fewer successful implementations of a complete JIT production system. We will return to this subject in Chapters 9, 10, and 17.

MANUFACTURING OUTPUTS PROVIDED BY THE FMS PRODUCTION SYSTEM

An FMS production system is a line flow production system designed to produce a wider variety of products in lower volumes than any other line flow production system (including operator-paced, JIT, equipment-paced, and continuous flow). The FMS production system uses expensive, flexible equipment and expensive computer hardware and software to form a physical line where many products can be produced in low volumes. The flexible equipment and computer software permit fast setups, continuous monitoring of quality, and rapid processing of very small batches of parts.

Because FMS is a line flow production system, it can provide the cost, quality, and delivery of a line flow production system. It also provides a high level of flexibility because it produces many products in low volumes. However, it cannot provide innovativeness at the same high level. Introducing new products and making design changes to existing products is difficult because these tasks require major changes to computer programs, fixtures, and so on.

SITUATION 2.5

When FMS Is the Only Choice

AN FMS at the Perkins diesel engine plant in the U.K. produced 50 different products in volumes that averaged 13 units per day, per product. The Pratt and Whitney FMS in Halifax, Nova Scotia, produced 70 different parts in volumes ranging from 30 to 1,000 units per year, per product. These numbers of products and volumes would normally be produced on a job shop or batch flow production system. When very high levels of quality and delivery are required, the FMS is the only production system that can be used. Neither the traditional line flow production systems—operator-paced, equipment-paced, and continuous flow—nor the new JIT production system can be used for this mix and volume of products.

Summary

There are seven different production systems:

- Job shop production system
- Batch flow production system
- Operator-paced line flow production system
- Equipment-paced line flow production system
- Continuous flow production system
- Just-in-time (JIT) production system
- Flexible manufacturing system (FMS) production system

Each is able to provide a unique set of the cost, quality, performance, delivery, flexibility, and innovativeness manufacturing outputs. And each system is uniquely suited to produce a particular mix of products and volumes. One of the tasks of manufacturing strategy is to select the best production system for each product or product family. How this is done is described in the next few chapters.

Further Reading

Dertouzos, M., R. Lester, and R. Solow, *Made in America: Regaining the Productive Edge*, New York: Harper Perennial Press, 1990.

Hayes, R., and S. Wheelwright, "Link Manufacturing Process and Product Life Cycles," *Harvard Business Review*, January–February, 1979.

Schmenner, R., *Plant and Service Tours in Operations Management*, New York: Macmillan Publishing Co., 1989.

Womack, J., D. Jones, and D. Roos, *The Machine That Changed the World*, New York: Harper Perennial Press, 1991.

CHAPTER 3

Manufacturing Levers: Designing and Changing Production Systems

Effective manufacturers find it useful to divide a production system into six subsystems (see Figure 3-1):

- Human resources
- Organization structure and controls
- Sourcing
- Production planning and control
- Process technology
- Facilities

These subsystems are called *manufacturing levers* to reflect the notion that each subsystem can be adjusted. Adjustments, which are usually made in response to changes in the external environment, are of many types. Small adjustments can be made to one or more levers to improve the current production system. Any new manufacturing technique or technology (see Chapter 10) can be thought of as a group of adjustments to several levers. Extensive adjustments can be made to all six levers for the purpose of changing from the current production system to a different system. An adjustment to a manufacturing lever is a change to a subsystem that is brought about by a management

FIGURE 3–1

Manufacturing Levers: The Six Subsystems that Make Up a Production System

Human resources	The skill level, wage, training and promotion policies, employment security, and so on, for all groups of employees.
Organization structure and controls	The formal relationships between groups (line and staff) in the production system. How are decisions made? What is the underlying culture? What systems are used to measure performance and provide incentives?
Sourcing	The amount of vertical integration. How does the production system manage those parts of the production and distribution system that it does not own? What is its relationship with suppliers?
Production planning and control	The rules and systems that plan and control: • the flow of material • the activities of line personnel • production support operations • the introduction of new products
Process technology	The nature of the production processes, the type of equipment, the amount of automation, and the linkages between the parts of the production process.
Facilities	The location, size, and focus of individual plants. The types and timing of changes to these plants.

decision. The positions of the six levers (that is, the decisions made in the six subsystems) completely determine:

- The type of production system
- How well the production system works
- The levels at which the manufacturing outputs will be provided

In this chapter we describe the manufacturing levers, how they interact with each other, and how they form a production system. Other ways of dividing a production system into subsystems have been proposed. While some have more subsystems and some have fewer, all are similar. A brief overview of these approaches is given in the appendix at the end of this chapter.

The Six Manufacturing Levers

Human Resources

The human resources subsystem comprises the organization's policies about its employees in the production system under consideration. The decisions made in this subsystem include:

- The mix of skilled and unskilled employees
- The number of job classifications
- Whether employees will be multiskilled
- The amount of training done
- The level of supervision
- The policy regarding layoffs
- The amount of responsibility and decision making given to employees
- The participation of employees in problem solving and improvement activities
- Promotion opportunities

Organization Structure and Controls

This subsystem comprises the organizational structure, reward systems, and culture. Decisions include the following:

- Whether the organizational structure is hierarchical or flat
- The relative importance of line and staff departments, and the use of teams and committees
- The amount of responsibility and authority at each level in the organization
- The measures used to evaluate the performance of individuals and departments
- Who is responsible for quality
- How managers are selected
- Whether the production system is a cost or profit center

Sourcing

The sourcing lever focuses on relationships with suppliers and distributors. Decisions made in this subsystem include:

- The number of suppliers and their capabilities
- Whether the relationship with suppliers is adversarial or a partnership
- The responsibility given to suppliers for design, cost, and quality
- The procedure for deciding whether a part will be produced internally or obtained from a supplier

Production Planning and Control

This subsystem consists of order entry, master production scheduling, materials planning, scheduling of machines and employees, controlling production on the plant floor, coordinating production support departments, and so on. Decisions that are made include:

- Whether systems are centralized or decentralized
- The size of raw material, work-in-process, and finished goods inventories
- Whether a push or pull production control system is used
- Whether frozen schedules are used
- When maintenance is done
- How design changes and new product introductions are scheduled into production

Process Technology

The process technology subsystem consists of the machines, processes, and technologies used to produce the products in the production system. Typical process technology decisions are:

- Determining the plant layout
- Whether machines are general purpose or specialized
- Whether tooling is low volume or high volume

- The amount of automation
- Whether technology is developed internally or purchased from external sources
- Whether the layout and technology are static or are the focus of continuous improvement efforts
- The procedures for quality control

Facilities

The facilities subsystem includes the buildings within which production takes place and the production support departments such as material handling, maintenance, engineering, and tooling. Decisions that are made include:

- Whether facilities are large or small
- Whether facilities are general purpose or special purpose
- The location of the facilities
- Capacity planning
- Improving the capabilities of production support departments

Interactions Among Manufacturing Levers

The six manufacturing levers constitute a production system; that is, the arrangement or pattern of the six levers completely determines whether the production system is a job shop, batch flow, operator-paced line flow, equipment-paced line flow, continuous flow, just-in-time, or FMS. Consequently adjustments to the manufacturing levers should not be made in a haphazard way. When considering a possible adjustment to a manufacturing lever, three checks should be made.

1. Is the adjustment appropriate for the production system?

The decisions made in each of the six subsystems must be appropriate for the particular production system used. For example, human resources policies that are appropriate for a job shop production system are not appropriate for a just-in-time production system. New techniques and technologies can be viewed as sets of adjustments to manufacturing levers. Consequently, of the many techniques and technologies available to a manufacturer, only those that are appropriate for the production system in use should be considered. For instance, suppose a manufacturer with a job shop production system hires a new manager whose previous experience was as a manager on an equipment-paced line flow production system. Any changes proposed by the new manager that draw on previous work experience should be checked to ensure that they are appropriate for the job shop production system.

2. Will the adjustment help provide the required manufacturing outputs?

Adjustments to a lever affect the manufacturing outputs provided by the production system. Of the many adjustments that can be made, only those that most help the production system provide the required outputs should be selected. To illustrate, suppose that flexibility and innovativeness form the basis of competition for a manufacturer with a batch flow production system. The manufacturer is considering possible changes in the human resources area. Of the many adjustments that can be made to this lever, those leading to an increase in the levels of flexibility and innovativeness should receive highest priority. For example, the incentive wage scheme could be adjusted to encourage operators to do rapid setups and produce products in smaller batches. Those who avoid setups by producing in large batches could be penalized.

3. How will the adjustment affect the other levers?

When making a decision in one subsystem, the implications of that decision on the other five subsystems should be considered. For example, changing the production planning and control system may require changes in human resource policies (such as retraining and changing job descriptions), in sourcing (such as improving supplier deliveries), and in organization structure and controls (such as adjusting reporting procedures).

Figure 3-2 illustrates the relationship among the production system, the manufacturing outputs, and the manufacturing levers. When the six levers are set in the appropriate positions for the production system in use, the production system is said to be well managed and the desired manufacturing outputs will be provided at the highest possible levels.

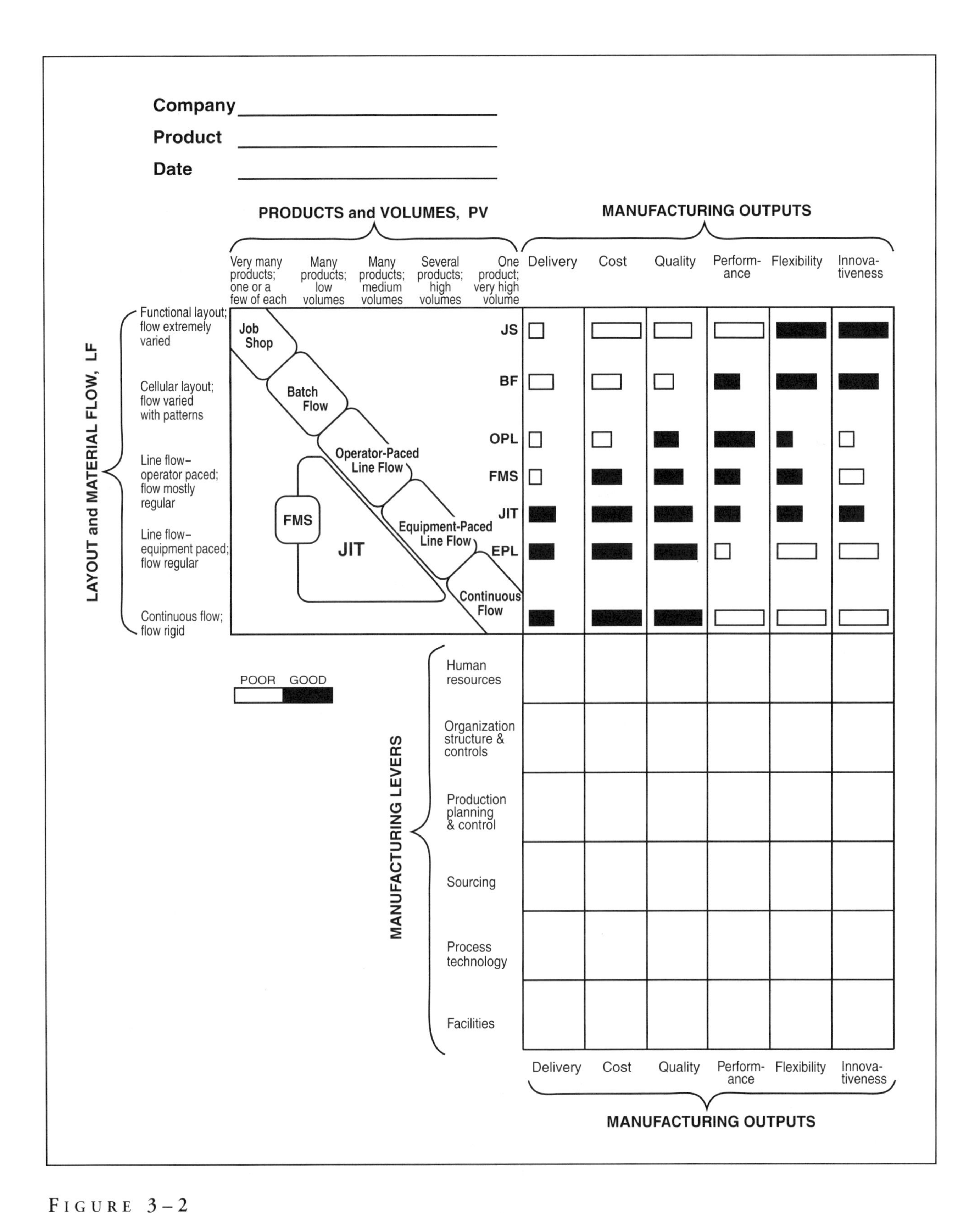

FIGURE 3–2

The Relationship between Production Systems, Manufacturing Outputs, and Manufacturing Levers

TRADE-OFFS

It is not difficult to determine the current settings of the manufacturing levers in an actual production system. A walk through the plant to note the types of equipment, layout, material handling procedures, routines of operators and maintenance personnel, attitudes of managers and staff specialists, planning procedures and information systems, and so on, gives a good picture of how each lever is set.

When faced with choices about possible adjustments to a lever, many manufacturing managers state that they simply try "to make the best decision." This implies a belief that it is possible to have the best of everything, that there are no trade-offs. This belief is mistaken. There is no best decision for all situations. Trade-offs must be made. For example, training employees to be multiskilled increases cost because of training expenses and higher wages, but it also increases flexibility and innovativeness. Production planning and control systems can be designed to minimize delivery time, or setup cost, or finished goods inventory. Facilities may consist of a few large, general-purpose plants or many small, specialized plants. Different incentive wage plans can be designed, each of which encourages different behavior. Products can be designed to maximize performance or to minimize cost.

SITUATION 3.1

Prestige Furniture Inadvertently Changes Production Systems

THE PROBLEM

The Prestige Furniture Company manufactured high-quality furniture to customer specifications. A customer might bring in an antique furniture piece and ask that a reproduction be made of it. Another customer might ask for a copy of a furniture set that was used during the reign of Louis XIV. Furniture was manufactured from customer drawings or from drawings prepared by the company's designers. The drawings were released to shop supervisors, who made all the production decisions. All manufacturing was done in a 30-year-old plant on a job shop production system.

A few years ago, two identical sets of bedroom furniture were produced by accident. No one knew what to do with the extra set. It was sold finally to a leading out-of-town furniture store for a relatively high price. Soon thereafter, it became a policy to produce one or more duplicate sets of furniture for each order. All duplicate sets were sold to out-of-town dealers, so customers never knew that somewhere there were duplicate sets of their furniture.

The new policy was very profitable. However, work in the shop became increasingly difficult. Many departments could barely handle the higher workload. Frequently parts became mixed up and some were even lost. Prestige Furniture realized that it needed to make some changes. A few high-speed, mass-production machines were purchased to give the plant extra capacity. Many simpler, more labor-intensive machines were removed to make room for the new machines. While everyone agreed that more changes were needed, including some sort of identification system for the parts, many employees were uncomfortable with the prospect of new machines, new systems, new skills, and new employees. Their attitude was, "If we did things well in the past without gimmicks, we can do them well now too."

Analysis

Before the policy of producing duplicate sets began, Prestige Furniture had a well-managed job shop production system (see Figure 3-3). The new policy increased production volumes, which moved Prestige horizontally away from the diagonal of the PV–LF matrix. An adjustment was made to the process technology lever (new equipment was acquired). An adjustment was planned for the production planning and control lever (a new part identification scheme). These adjustments would begin to change the job shop production system to a batch flow production system. Other adjustments would then need to be made to all six levers so that the new production system would operate properly. Some of these adjustments are shown in the manufacturing levers section of Figure 3-3. It appeared that the adjustments to the

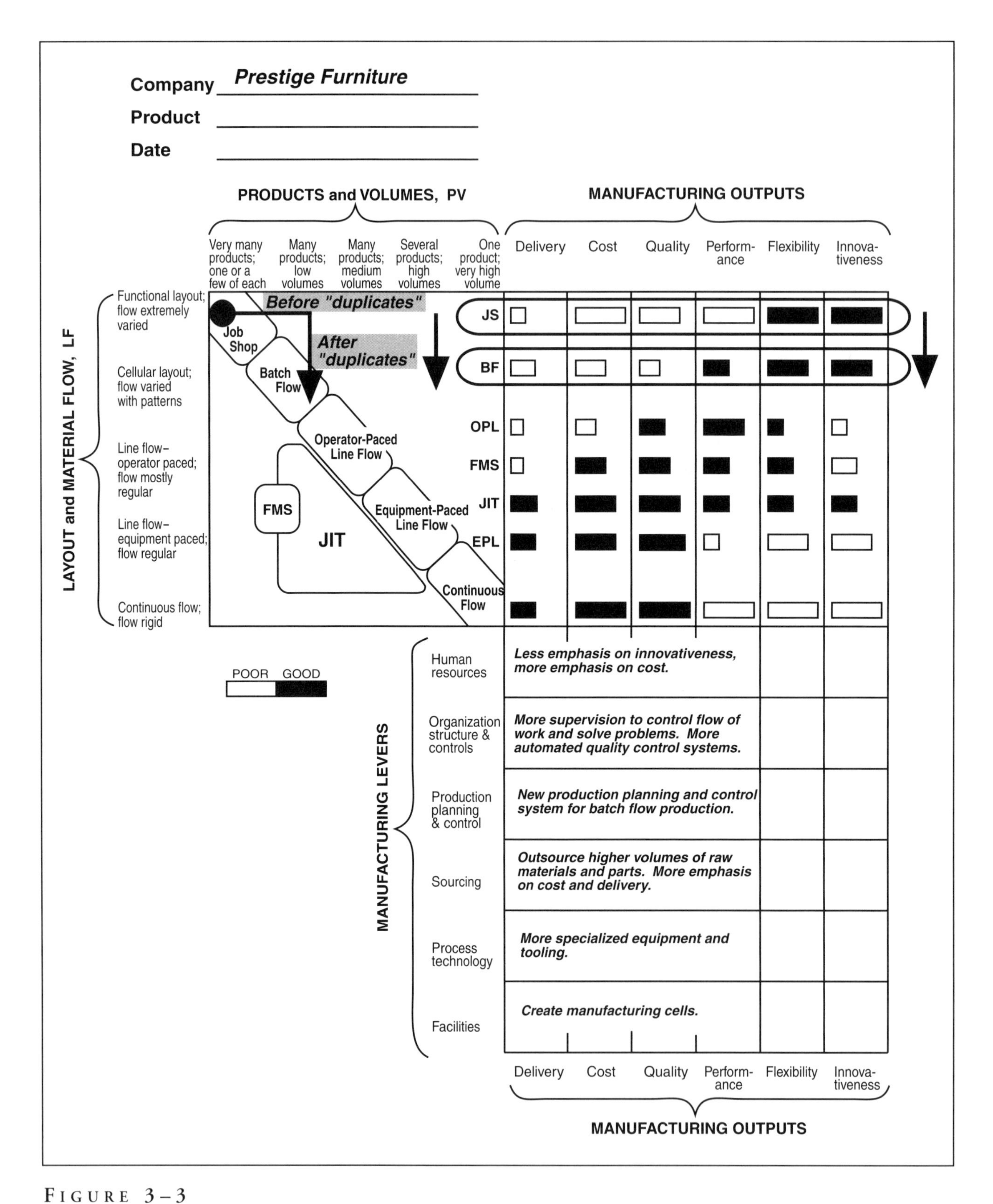

FIGURE 3–3

Analysis of Prestige Furniture's Implicit Decision to Change Production Systems

human resources and production planning and control levers would be particularly difficult.

In fact, Prestige never made the adjustments to the manufacturing levers that were needed to produce a proper batch flow production system. They continued to have production problems, and management spent most of its time "putting out fires." The production system never worked smoothly. The levels at which the manufacturing outputs were provided were lower than they should have been. This situation continued because Prestige Furniture had no competitors. The company was quite vulnerable to any competitor with a well-managed job shop or batch flow production system.

Prestige should never have changed its job shop production system. They did not have sufficient manufacturing capability to change from one production system to another (see Chapter 5). They should have set up a separate plant-within-the-plant to produce small quantities of selected products on either a batch flow or operator-paced line flow production system. This would have satisfied their desire for more business and given them two good production systems instead of one poor one.

SITUATION 3.2

COR Shirts Changes from Batch to Line Flow

PART I
THE PROBLEM

From humble beginnings, COR Shirts became a fair-sized producer of men's shirts. According to Mr. Walker, the company's general manager, "Competition in the shirt business is cutthroat. It is essential to keep production costs down, despite increasing wage rates and rising materials costs. It is also important to maintain quality of cloth and workmanship."

The plant had a functional layout. Cloth was cut in the cutting department, stitching was done in the large sewing department, and pressing and packaging were done in the packaging department. Whenever the inventory of a product (that is, a size, style, and color of shirt) fell below two dozen shirts, a clerk wrote a production order to produce ten dozen shirts. Production orders for new products were written by the production manager. New products were designed on the recommendation of sales personnel, who were informed by distributors about the popularity of new styles. COR could design, produce, and supply any new style within three days. Products were always produced in bundles of ten dozen identical shirts. These bundles moved from department to department, and from workstation to workstation, until all the operations were completed.

ANALYSIS

COR used a batch flow production system. The company's competitive advantage was in providing high levels of flexibility and innovativeness. COR had to produce many sizes, styles, and colors of shirts. It also had to be able to introduce new shirts quickly. The levels of the cost and quality outputs were of concern to the general manager. However, it may not be possible to provide these outputs at the levels required by the general manager with this production system (see Figure 3-4).

PART 2

THE PROBLEM

Mr. Walker decided to change the production system to reduce costs. A new line layout was implemented, patterned after a television assembly line that Mr. Walker had recently visited. In the first few weeks after the implementation, the production flow was not as smooth as originally anticipated and bottlenecks were apparent at several workstations. Most of the problems disappeared when the size of the bundles was increased and a new procedure for releasing production orders was established.

The new production system was excellent in terms of efficiency, and the plant would have operated with extremely low unit costs had sales been large enough to use the plant's increased capacity. But sales were declining instead of increasing. It soon became apparent that customers were reducing their orders. Panicky inquiries revealed that in the "standard shirt" business, COR was just one of many shirt companies. Customers had bought from COR Shirts because of its large variety, originality, and style.

ANALYSIS

In his attempt to reduce costs, Mr. Walker changed the production system to a line flow production system. This required higher volumes of fewer products (see Figure 3-4). With the new production system came

FIGURE 3–4

Analysis of COR Shirts

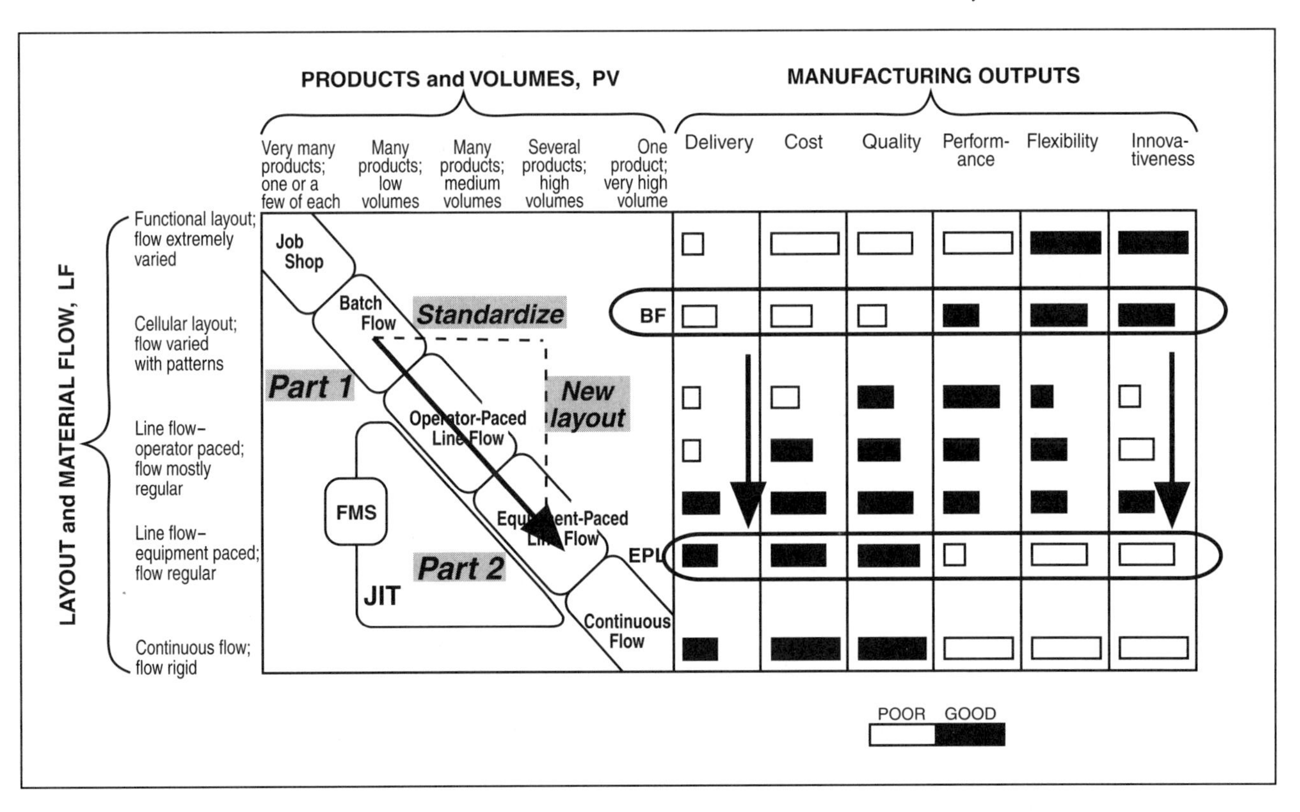

improved levels of cost and quality. However, the levels of flexibility and innovativeness dropped and, since this was what customers wanted, orders decreased.

At this point Mr. Walker had two choices. If he kept the line flow production system, COR Shirts would have to look for new customers and compete in the standard shirt business. He could keep his old customers if he changed the production system back to a batch flow system, which is what he did. The line was dismantled and the equipment was rearranged to form a batch flow production system. Mr. Walker should never have changed the existing production system. He should have made improvements, using one of the improvement approaches to be discussed in Chapter 9, to raise the level of the cost and quality outputs. A small plant-within-the-plant could also have been organized to produce a small number of standard shirts in high volumes on a line flow production system.

Summary

Effective manufacturers find it useful to partition a production system into six subsystems:

- Human resources
- Organization structure and controls
- Sourcing
- Production planning and control
- Process technology
- Facilities

The six subsystems are called the manufacturing levers. The arrangement or pattern of the six levers completely determines whether the production system is a job shop, batch flow, operator-paced line flow, equipment-paced line flow, continuous flow, just-in-time, or FMS. Consequently, adjustments to the manufacturing levers should not be made haphazardly. Each adjustment should be appropriate for the production system used and should help the production system provide the manufacturing outputs at the required levels. Small adjustments can be made to one or more levers to fine-tune an existing production system. Groups of adjustments represent improvement approaches and new manufacturing techniques and technologies. Extensive adjustments to all six levers are required when the existing production system is changed to a different one.

Appendix: Defining the Subsystems That Constitute a Production System

Many authors have developed lists of strategic decision areas or subsystems within manufacturing. Each list shares the following properties:

- The list is comprehensive; that is, all manufacturing decisions fall within the decision areas on the list.
- The list is discriminating. It is possible to break complex manufacturing decisions into analyzable pieces, each of which falls neatly within specific decision areas.
- The list is reflective. The decision areas are consistent with manufacturing's view of itself.

Five well-known lists with these properties are shown in Figure 3-5. Studies have shown that there is a close match between these lists and the decision areas used by practicing manufacturing managers (Schroeder, Anderson, and Cleveland 1986; Ward, Miller, and Vollmann 1988).

The decision areas are partitioned into two groups, structural decision areas and infrastructural decision areas, as suggested by Hayes and Wheelwright (1984). They liken structural decision areas to computer hardware, and infrastructural decision areas to computer software. The analogy leads to the following observations:

- Structural and infrastructural decision areas are of equal importance.
- Both are dependent on each other.
- Manufacturers who ignore either decision area for long will not be successful.

Type of Decision Area	Skinner (1974)	Buffa (1984)	Hayes, Wheelwright, and Clark (1984, 1988)	Fine and Hax (1985)	This Book
Structural	• Plant and equipment	• Capacity and location • Product and process technology • Strategy with respect to suppliers and vertical integration	• Capacity • Facilities • Technology • Vertical integration	• Capacity • Facilities • Processes and technologies	• Facilities • Process technology • Sourcing (suppliers and vertical integration)
Infrastructural	• Production planning and control • Organization and management • Labor and staffing • Product design and engineering	• Strategic implications of operating decisions • Workforce and job design • Position of production system	• Production planning and control • Quality • Organization • Workforce • New product development • Performance measurement systems	• Product quality • Human resources • Scope of new products	• Production planning and control • Organization structure and controls • Human resources

Source: Adapted from Leong, Snyder, and Ward 1990.

FIGURE 3–5

Defining the Subsystems that Constitute a Production System

REFERENCES

Buffa, E. S., *Meeting the Competitive Challenge*, New York: Dow Jones-Irwin, 1984.

Fine, C., and A. Hax, "Manufacturing Strategy: A Methodology and an Illustration," *Interfaces*, Vol. 15, No. 6, pp. 15–27, 1985.

Hayes, R. H., and S. C. Wheelwright, *Restoring Our Competitive Edge: Competing Through Manufacturing*, New York: John Wiley & Sons, 1984.

Hayes, R. H., S. C. Wheelwright, and K. B. Clark, *Dynamic Manufacturing: Creating the Learning Organization*, New York: Free Press, 1988.

Leong, G. K., D. L. Snyder, and P. T. Ward, "Research in the Process and Content of Manufacturing Strategy," *Omega: The International Journal of Management Science*, Vol. 18, No. 2, pp. 109–122, 1990.

Schroeder, R. G., J. Anderson, and G. Cleveland, "The Content of Manufacturing Strategy: An Empirical Study," *Journal of Operations Management*, Vol.6, No.4, pp.405–415, 1986.

Skinner, W., "The Focused Factory," *Harvard Business Review*, pp. 112–121, May–June 1974.

Ward, P., J. G. Miller, and T. Vollmann, "Mapping Manufacturers' Concerns and Action Plans," *International Journal of Operations and Production Management*, Vol. 8, No. 6, pp. 5–17, 1988.

CHAPTER 4

Competitive Analysis: Selecting the Best Production System

More than 20 years ago, Wickham Skinner (1969) told the story of a company called Electronic Gear, which produced five kinds of electronic products for three different markets. One market demanded extremely high quality. Rapid introduction of new products was demanded in the second market. In the third, low cost was critical. In spite of these diverse and contrasting requirements, management at Electronic Gear had centralized all manufacturing activities in one plant to achieve economies of scale. The result was a failure to achieve high quality, low cost, or the ability to introduce new products quickly. Manufacturing satisfied none of its customers, serious marketing problems occurred, and the company struggled. More than two decades later, there are still many companies like Electronic Gear. The company should have organized plants-within-the-plant, each with a different production system capable of providing the particular outputs desired by the customers.

What is the mechanism for deciding which of the seven production systems is most suitable for a particular situation? The *competitive analysis* element of manufacturing strategy is where this is sorted out. Customer requirements are translated into specific manufacturing outputs, target levels are set for each output, and a suitable production system is determined. Several attempts or iterations are usually required to successfully complete this element of the manufacturing strategy.

Performing a Competitive Analysis

The first task for any manufacturer is to identify customer needs and expectations. This is not easy because there are six manufacturing outputs and each output has a number of dimensions or attributes. The manufacturer must identify the key outputs and attributes that make up customer needs and expectations. A short list of attributes is given in Figure 4-1. (See also Figure 1-4 on page 25.) Time, effort, and resources are necessary for performing a proper competitive analysis. A simple and effective procedure consists of the following five steps:

Step 1: Attributes are defined for each manufacturing output. Some frequently used attributes are listed in Figure 4-1.

Step 2: Numerical estimates are obtained. Data is collected on each attribute for:

- The company's product
- A typical (or average) product in the marketplace
- A strong competitor's product

Step 3: Each manufacturing output is classified as one of the following (these terms are defined below):

- A market qualifying output
- An order winning output
- An unimportant output

Step 4: Twelve-month targets are set for all attributes of the market qualifying and order winning outputs.

Step 5: The best production system is selected. (If necessary, return to step 3.)

Steps 1, 2, and 4 involve extensive collection and analysis of data from inside and outside the company. Benchmarking is the tool used to do this. (See Chapter 9.)

Terry Hill (1989) introduced the concept of market qualifying and order winning outputs. Their definitions follow from an observation made in Chapter 3: *Manufacturing can provide some, but not all, outputs at very high levels.* Each organization needs to

Cost	Unit product cost Capacity/machine utilization Inventory turnover Process yield Labor productivity
Quality	Percent defective Scrap and rework costs Warranty costs Quality of incoming material from suppliers
Performance	Number of standard features Number of advanced features Mean time between failures
Delivery	Quoted delivery time Percent on-time deliveries Order entry cycle time Average lateness
Flexibility	Number of products in product line Number of options Minimum order size Average lot size Length of frozen schedule
Innovativeness	Lead time to design new products Lead time to prepare customer drawings Number of engineering change orders per year Number of new products introduced each year

FIGURE 4–1

Frequently Used Attributes for Each Manufacturing Output

determine the outputs that it will provide at the highest possible levels and those that it will provide at somewhat lower levels.

MARKET QUALIFYING OUTPUTS

Market qualifying outputs are those outputs that customers expect to receive. A product needs these outputs to compete in the marketplace. If manufacturing does not provide them, the company is at a great disadvantage. Providing a market qualifying output means providing each attribute of that output at a very high level, called the market qualifying level.

ORDER WINNING OUTPUTS

An order winning output is an output that is not common in the marketplace. It can be considered a market qualifying output provided at a level that is much higher than the market qualifying level—the order winning level. Order winning outputs differentiate manufacturers from one another. They are the reasons why customers buy from particular manufacturers. If the level of an order winning output can be raised, then more orders will result. Providing an output at an order winning level is equivalent to being among the best in the world for that output for the particular product produced.

Recall Figure 1-2 (on page 17), where examples were given of products for which one manufacturing output was more important than the others. These outputs are the order winning outputs. For example, performance was the order winning output for Tide detergent. Customers buy Tide because of its uniquely high level of performance. Competitors' products cannot clean clothes as well as Tide can. Cost and quality are also important manufacturing outputs, and Procter & Gamble, the manufacturer of Tide, provides these outputs at high, market qualifying levels. Performance is the order winning output, however, so it is provided at a much higher level than the market qualifying level. It is provided at the highest possible level in the world, a level that few competitors can match. The first task of manufacturing is to provide the market qualifying outputs at market qualifying levels. When this is done, manufacturing strives to provide an order winning output at its order winning level.

The levels at which manufacturing outputs are provided depend on two factors:

- The production system used
- The level of manufacturing capability of the production system (see Chapter 5)

A fascinating illustration of the implications of these concepts is the incident that occurred between Honda and Yamaha in the early 1980s (see Situation 4.1).[1]

Honda raised the level of its market qualifying outputs, cost and delivery, and its order winning outputs, performance and innovativeness, to such an extent that it almost forced its competitor, Yamaha, out of business. Honda could do this because it

1. The Honda-Yamaha War is described in G. Stalk, "Time—The Next Source of Competitive Advantage," *Harvard Business Review*, pp. 41–51, July–August 1988.

SITUATION 4.1

The Honda-Yamaha War

THE HONDA-YAMAHA war started in 1981, when Yamaha announced the opening of its new motorcycle factory. The factory made Yamaha the world's largest motorcycle manufacturer, a prestigious position previously held by Honda. Honda had been concentrating its resources on its automobile rather than its motorcycle business, but now, faced with Yamaha's public challenge, it chose to counterattack.

In the no-holds-barred war that followed, Honda provided its customers with higher and higher levels of its manufacturing outputs. Honda raised the levels of its market qualifying outputs, cost and delivery, by cutting prices and flooding distribution channels. It raised the levels of its order winning outputs, innovativeness and performance, by introducing new products and raising the technological sophistication of all its products. At the start of the war, Honda and Yamaha each had 60 models of motorcycles. Over the next 18 months, Honda introduced or replaced 113 models, while Yamaha was only able to manage 37 changes. Honda also introduced four-valve engines, composite materials, direct drive, and other features.

Yamaha could not provide these manufacturing outputs at the new market qualifying levels, let alone at order winning levels. Demand for Yamaha products disappeared. At the most intense point in the war, Yamaha had more than 12 months of inventory in its dealer showrooms.

Finally Yamaha surrendered. In a public statement, Yamaha President Eguchi announced, "We want to end the Honda-Yamaha war. It is our fault. Of course, there will be competition in the future, but it will be based on a mutual recognition of our competitive positions."

used production systems that were better able to provide these outputs than the system that Yamaha used. Honda used operator-paced line flow production systems and just-in-time production systems, while Yamaha used an equipment-paced line flow pro-

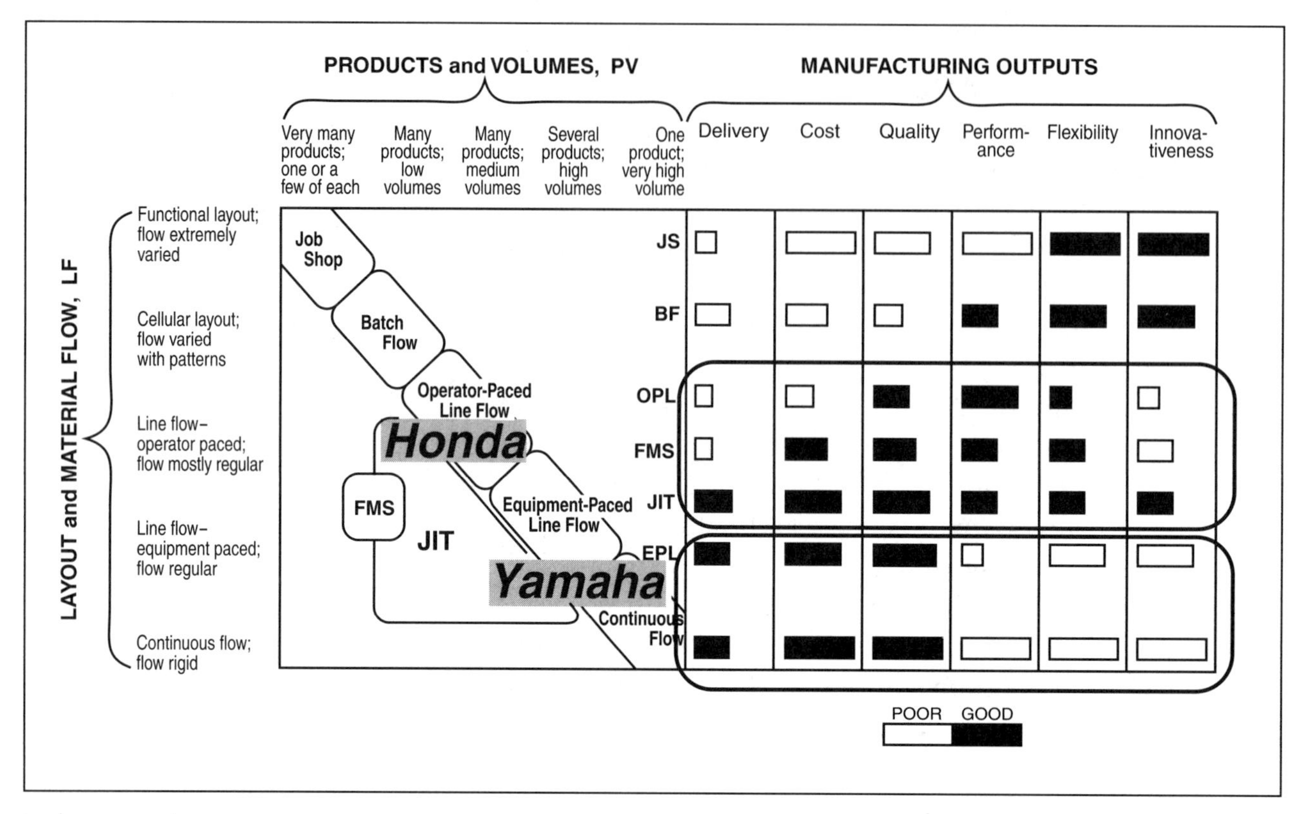

FIGURE 4–2

Production Systems and Manufacturing Outputs in the Honda-Yamaha War

duction system (see Figure 4-2). In addition, Honda's older production systems had higher levels of manufacturing capability than Yamaha's new production system (see Chapter 5).

The relationship between the five-step process for doing a competitive analysis and the other elements of manufacturing strategy is illustrated in Figure 4-3. This framework is used by multifunctional teams (with members from manufacturing, marketing, human resources, finance, and so on). How they determine the market qualifying and order winning outputs for each product (or product family), the production system, and the adjustments to the manufacturing levers is discussed in the next section.

A Competitive Analysis at ABC Company

An example, adapted from an actual problem at a Fortune 500 company, illustrates the elements of manufacturing strategy presented so far. A plant at ABC Company manufactured four product families. Three product families were built to customer order and required significant amounts of custom engineering.

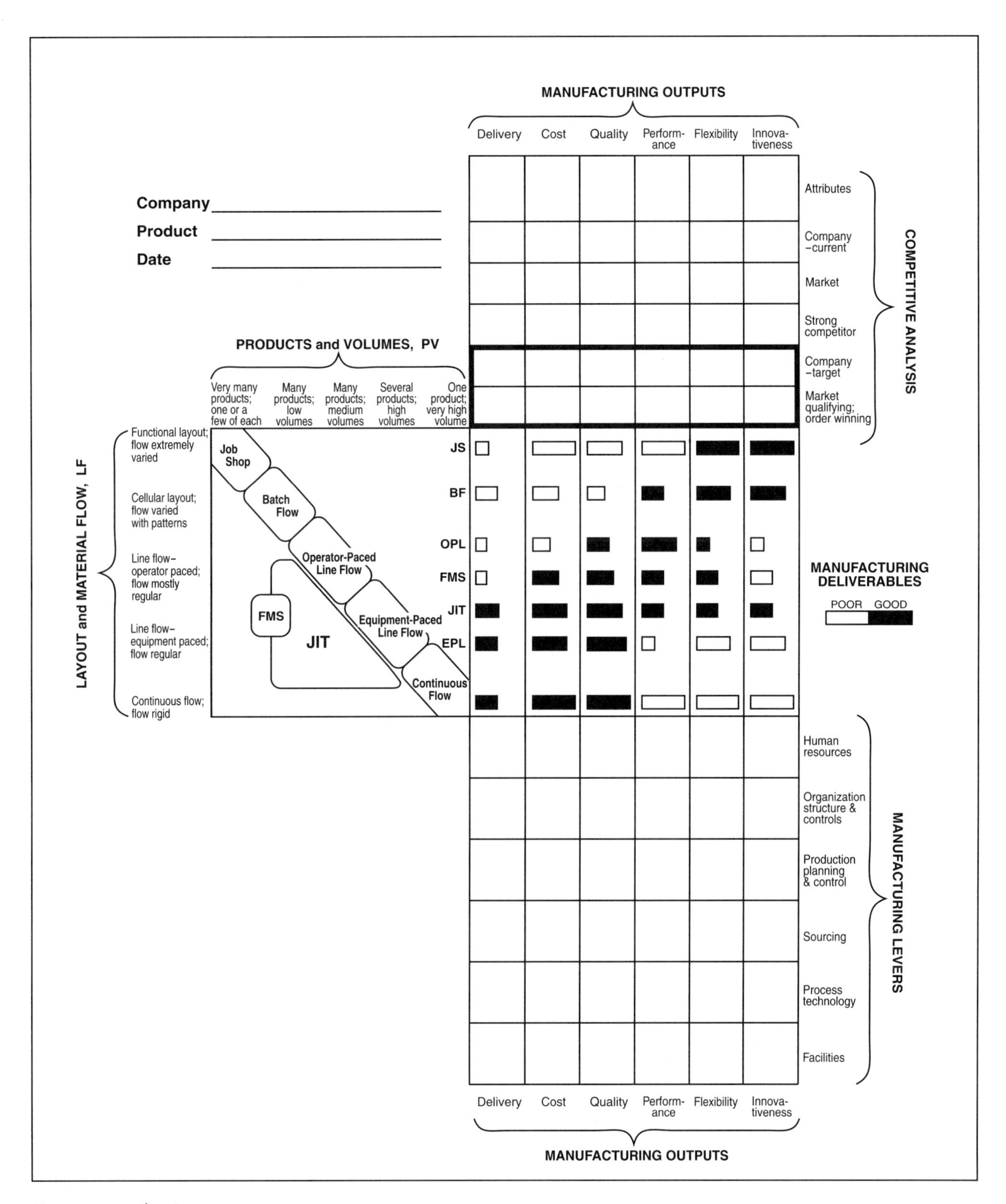

FIGURE 4–3
Adding the Competitive Analysis

The other product family consisted of standard products. All products were manufactured in a modern plant using a batch flow production system. The plant had been unprofitable for most of the last ten years. Top corporate management decided to make one last attempt to turn the plant around. They assigned a dynamic plant manager to the plant and told the new manager to make the plant profitable within four years, or they would close it down.

The plant manager put together a new management team. They quickly realized that improvements in the production of one product family, QH4500, would have to provide most of the profits for the plant. QH4500 had been recently redesigned and so was an up-to-date product. It had a sizable market share and accounted for almost 50 percent of the plant's sales. Before making any changes to the production system, the management team did the following competitive analysis.

Steps 1 and 2: Define Attributes and Collect Data

The management team asked the marketing department (located 100 miles away at corporate headquarters) for market data on cost, quality, performance, etc., levels of competitors' products. The marketing department was unable and unwilling to provide these data, and so the plant gathered it on their own. The important manufacturing outputs for QH4500 were quality, performance, delivery, and cost. Typical customer comments included the following:

- QH4500 is the "Cadillac" product in the marketplace,
- When you buy QH4500, you buy quality,
- Delivery is a problem in the marketplace. If ABC could provide the product in 16 weeks, it would own the market.

A summary of the important data from the competitive analysis is shown in Figure 4-4. Delivery time was the number of weeks between accepting an order and delivering it to the customer. It was currently 22 weeks. Delivery time reliability was the percentage of all customer orders that were delivered on time, which was currently 60 percent. Cost was defined as the factory cost of producing one QH4500 unit which was $40,000. Three major quality attributes were defined: rework costs, measured as

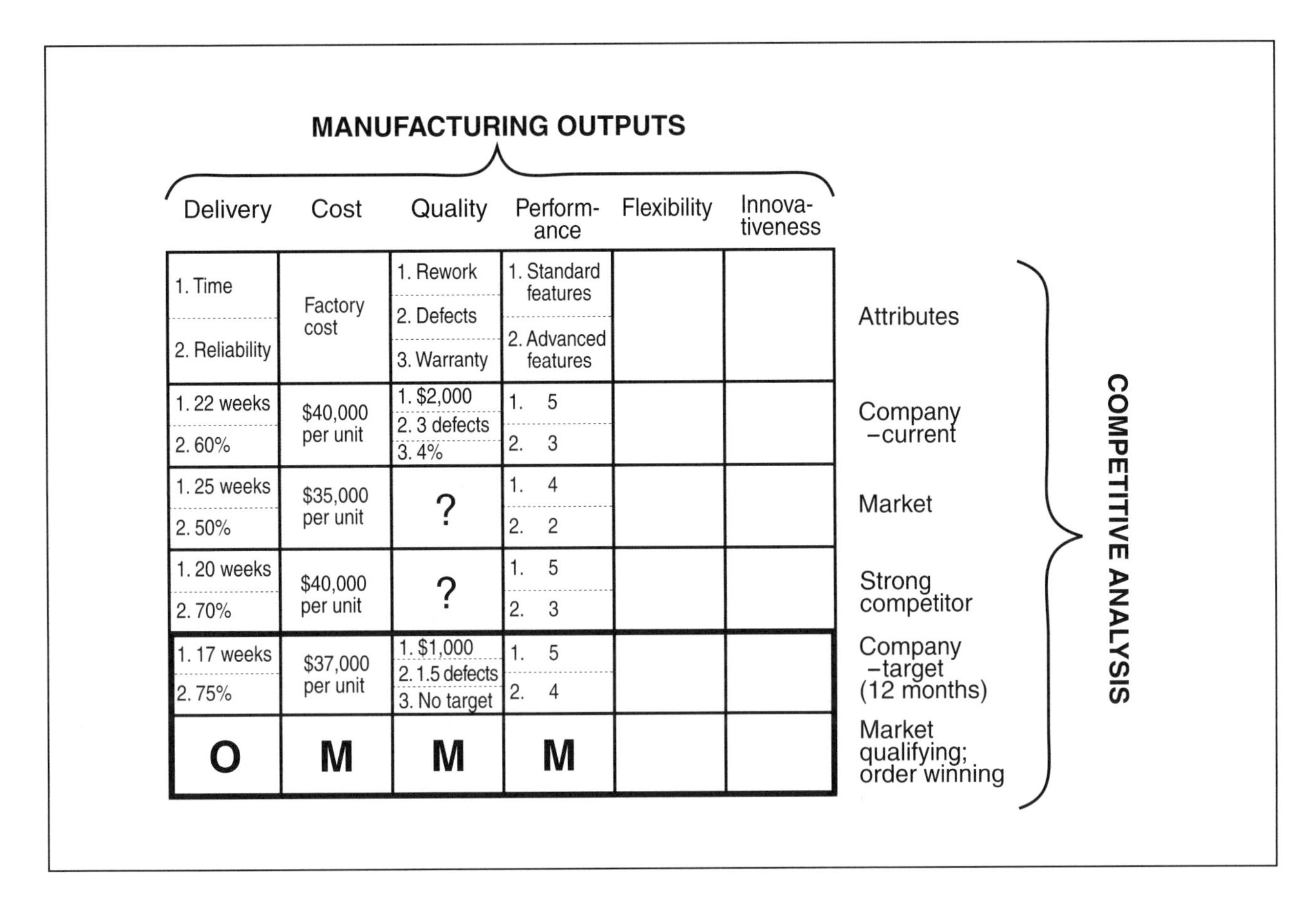

FIGURE 4–4

A Competitive Analysis for Product Family QH4500 at ABC Company

average factory rework dollars per unit (currently $2,000): average defects per unit detected in the final test department (currently three defects): and warranty costs as a percentage of sales (currently 4 percent). Performance consisted of two attributes: the number of standard features (currently five), and the number of advanced features (currently three).

The plant team also gathered data on competitors' products. From these data, the values shown in Figure 4-4 for the market's average product and for the product manufactured by the strongest competitor were determined. Notice that it was not possible to obtain values for all the attributes.

STEPS 3 AND 4: SELECT THE MARKET QUALIFYING AND ORDER WINNING OUTPUTS

The team decided that delivery, cost, quality, and performance were the market qualifying outputs. Twelve-month targets were set.

- The team decided that product cost had to be reduced to $37,000 per unit within 12 months for QH4500 to remain competitive in the marketplace.
- Improvements in quality were necessary. The target for rework cost was set at $1,000 per unit. A target of 1.5 defects per unit in the final test department was set. No target was set for warranty cost because it was expected to drop when improvements were made to the production system.
- To provide performance at a market qualifying level, an additional advanced feature would need to be added within the next 12 months.
- The market qualifying level for delivery was estimated to be a 20-week delivery time and 70 percent delivery time reliability.

Once these market qualifying targets were set, the team had to determine which of the market qualifying outputs would be the order winning output. Their deliberations went as follows.

Cost could not be the order winning output for two reasons. 1) The plant was located in a part of North America where the cost of doing business was too high to permit it to be the world's low-cost producer. 2) The volume of QH4500 was too low to permit the batch flow production system to be changed to an equipment-paced line flow production system, which would be necessary if ABC was to become the industry's low-cost producer. Quality was also not selected because the plant did not have the resources to make the quality improvements necessary to become the world leader in quality.

Performance seemed a good choice for the order winning output because QH4500 already had a reputation for excellent performance. There were two reasons, however, why this output was not selected. 1) One new advanced feature would need to be introduced in the next 12 months just to maintain performance at a market qualifying level. This new feature included a component made from a new composite material, which required new technology that was already straining the organization's engineering resources. The shortage of engineering resources made it impossible to provide performance at more than the market qualifying level. 2) There were communication problems between the plant and the product design group, which was located 100 miles away at corporate headquarters.

Delivery was the unanimous choice for the order winning output. Market research indicated that a delivery time near 17 weeks and a high delivery time reliability would result in increased orders. The plant was located near its North American customers, which gave it a natural advantage over its strong offshore competitors. The team felt that they had control over almost everything that affected delivery time. The order winning target level for delivery was set at 17 weeks, with a 75 percent delivery time reliability, both to be accomplished within 12 months.

The team was satisfied with its choice of market qualifying and order winning outputs. These were the outputs that customers wanted. Within 12 months, all outputs would be provided at much higher levels than the current levels. One of the outputs would be provided at the highest level in the world. The team then set out to determine how manufacturing would provide these outputs (more on this follows later in the chapter).

Competitive Analysis and the Production Systems

The levels at which the market qualifying and order winning outputs must be provided are always increasing. Every year customers expect lower costs, higher quality, better performance, better delivery, and more flexibility and innovativeness. Competitors provide ever-higher levels of these outputs as they improve their manufacturing capabilities. The competitive analysis determines the manufacturing outputs that must be provided by a production system. The production system that is most capable of providing the desired outputs at the required levels is the one that should be used. Sometimes this production system is the one currently used. In other cases, a different production system should be used. There is a range of possibilities:

- The current production system is the required production system.
- A new production system is required. It is feasible and can be achieved.
- A new production system is required. It is feasible but cannot be achieved.

- A new production system is required, but it is not feasible.
- There is no production system capable of providing the required outputs.

A production system is *feasible* for a particular situation when the number of products and the product volumes are appropriate for it. A new production system can be *achieved* when the organization's capabilities are sufficiently high that it can make the necessary adjustments to the manufacturing levers. (See Chapter 5.)

The Current Production System Is the Required Production System

In this case, the current production system is capable of providing the required market qualifying and order winning outputs at the target levels, provided the production system is well managed and has a sufficiently high level of manufacturing capability. Recall that "well managed" means that each manufacturing lever is set in the appropriate position for the particular production system.

A New Production System Is Required; It Is Feasible and Can Be Achieved

A different production system that is capable of providing the required market qualifying and order winning outputs at the target levels is identified. The number of products and the product volumes are appropriate for the new production system, and the organization has sufficient manufacturing capability to change from the current production system to the new one.

A New Production System Is Required; It Is Feasible but Cannot Be Achieved

The current production system is incapable of providing the required market qualifying and order winning outputs at the target levels. A different production system capable of providing these outputs is identified. It is feasible because the number of products and the product volumes are appropriate for the new production system, or they can be modified to make them appropriate. Careful study shows, however, that changing from the current production system to the new production system is impossible, given the organization's current capabilities. (See Chapter 5.) The new production system cannot be achieved.

When this occurs, the team investigates whether another feasible production system is capable of providing the required outputs, and whether this production system can be achieved. If none exists, the team must decide whether to raise the level of manufacturing capability, a long and difficult process (see Chapters 5 and 9), or to return to the competitive analysis and select a different set of market qualifying and order winning outputs.

A New Production System Is Required, but It Is Not Feasible

The current production system is not capable of providing the required market qualifying and order winning outputs at the target levels. A different production system capable of providing these outputs is identified. However, the new production system is not feasible because the current number of products and the product volumes are not compatible with the new production system, nor can they be modified to make them compatible. For example, a manufacturer with a job shop production system may wish to change to an equipment-paced line flow production system to achieve lower costs and faster delivery. However, the manufacturer may not have sufficient product volumes for using the dedicated lines required for this line flow system. Therefore, the equipment-paced line flow production system is not feasible.

When this occurs, the team developing the manufacturing strategy investigates whether a feasible production system is capable of providing the required outputs. If none exists, the team returns to the competitive analysis and selects a different set of market qualifying and order winning outputs.

There Is No Production System Capable of Providing the Required Outputs

The team returns to the competitive analysis and selects a different set of market qualifying and order winning outputs. Sometimes this occurs when a team decides that all outputs will be order winning. The target levels for these outputs are so high that it is impossible for any production system to provide them.

Competitive Analysis at ABC Company (Continued)

Step 5: Select the Best Production System

The competitive analysis done in Figure 4-4 is shown on the worksheet in Figure 4-5. Notice that the market qualifying and order winning outputs cannot be provided by the existing batch flow production system. (ABC used a batch flow production system to produce four product families in relatively low volumes.) A line flow production system—operator-paced, equipment-paced, JIT, or FMS—is required to provide these manufacturing outputs. The traditional operator-paced and equipment-paced line flow production systems are the most attractive systems. JIT is difficult to implement, and FMS is expensive.

Operator-paced and equipment-paced line flow production systems are feasible when a few standard products are produced in medium to high volumes. However, QH4500 was a family of custom-engineered products, with each product produced in low volumes. Three changes would be needed to make these production systems feasible for QH4500:

- The design group would need to standardize QH4500; that is, the number of options would have to be reduced, and QH4500 would need to be redesigned so that it would be easy to produce on a line flow production system.
- The marketing group would need to market a standard product with a limited number of options.
- The production volume of QH4500 would need to be increased so that the dedicated production line could be fully utilized.

Unfortunately, none of these changes could be made. The design group, which was located 100 miles away at corporate headquarters, was unwilling to participate in any standardization program beyond a small one to reduce the number of part numbers (which they were doing already as part of a project to improve their CAD system). The marketing group, which was also located at corporate headquarters, was opposed to any program that would reduce the options available to customers. As

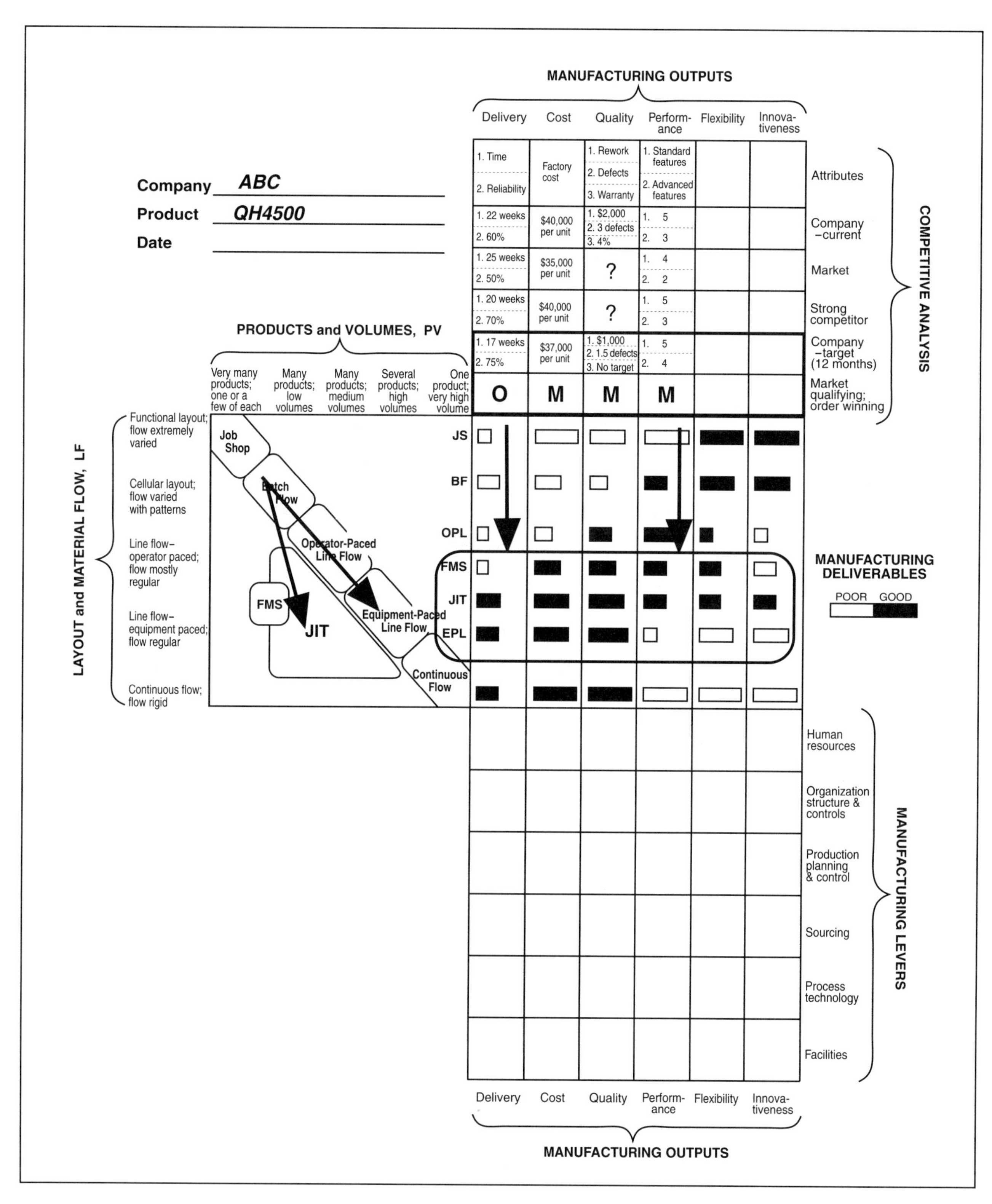

FIGURE 4–5

Selecting the Appropriate Production System for Product QH4500

for obtaining extra volume, the team was told that no new QH4500 volume would be given to the plant until it demonstrated that it could generate profits consistently.

The inability to standardize QH4500 and obtain higher volumes convinced the team that an operator-paced or equipment-paced line flow production system would not be feasible. The FMS production system was ruled out because the plant could not afford the expensive equipment required. Therefore, the team decided that the required production system for QH4500 was the difficult-to-implement JIT production system (see Figure 4-5).

Adjusting and Changing the Production System

When the current production system is the required production system, the manufacturing levers should already be set in the appropriate positions. All that is needed are small adjustments to fine-tune the system. Each manufacturing lever—human resources, organization structure and controls, sourcing, and so on—is checked to ensure that it is set in the appropriate position for the production system used. Then each lever is checked to determine the adjustments necessary so that the production system will be better able to provide the market qualifying and order winning outputs at the target levels.

When the current production system is changed to a new one, extensive adjustments are made to the manufacturing levers. Each lever is changed from its current position, which is appropriate for the current production system, to a new position that is appropriate for the new production system. More adjustments are then made so that the new production system will be better able to provide the required outputs at the target levels.

Making extensive adjustments to the manufacturing levers is difficult. Problems occur if it is not done carefully. An implementation plan is developed to organize the adjustments into a practical plan. It consists of:

- Identifying the adjustments that need to be made
- Determining the sequence in which the adjustments will be made
- Determining the pace at which the adjustments will be made

- Determining the resources that will be required, and so on

The implementation plan is the subject of Chapter 7.

Competitive Analysis at ABC Company (Continued)

Figure 4-6 shows some of the adjustments made to the manufacturing levers at the ABC plant when the batch flow production system was changed to a JIT production system. Major adjustments were made to the human resources lever. The workforce in a JIT production system is different from the workforce in a batch flow production system. Training programs were required to teach employees the skills needed to perform the new jobs in the JIT production system. Workers were taught to do a broad range of tasks. The plant and union negotiated a reduction in the number of job classifications and a new pay scheme that paid employees for the number of jobs they were qualified to perform rather than the specific job they did. These adjustments are represented in Figure 4-6 by the circled number 1 in the row corresponding to the human resources lever. The position of the circled number is significant for two reasons.

- *Row Position:* The circled number is placed in the row corresponding to the human resources lever because it represents an adjustment necessary to change this lever (or production subsystem) from one appropriate for the old production system to one appropriate for the new production system.
- *Column Position:* The circled number is placed in the columns corresponding to the delivery, cost, and quality manufacturing outputs to signify that, of all the adjustments that can be made to this lever, those that make it possible for the new production system to provide the required market qualifying and order winning outputs at the target levels are the ones that will receive the highest priority.

More details on the positions of levers in a JIT production system can be found in Chapter 17.

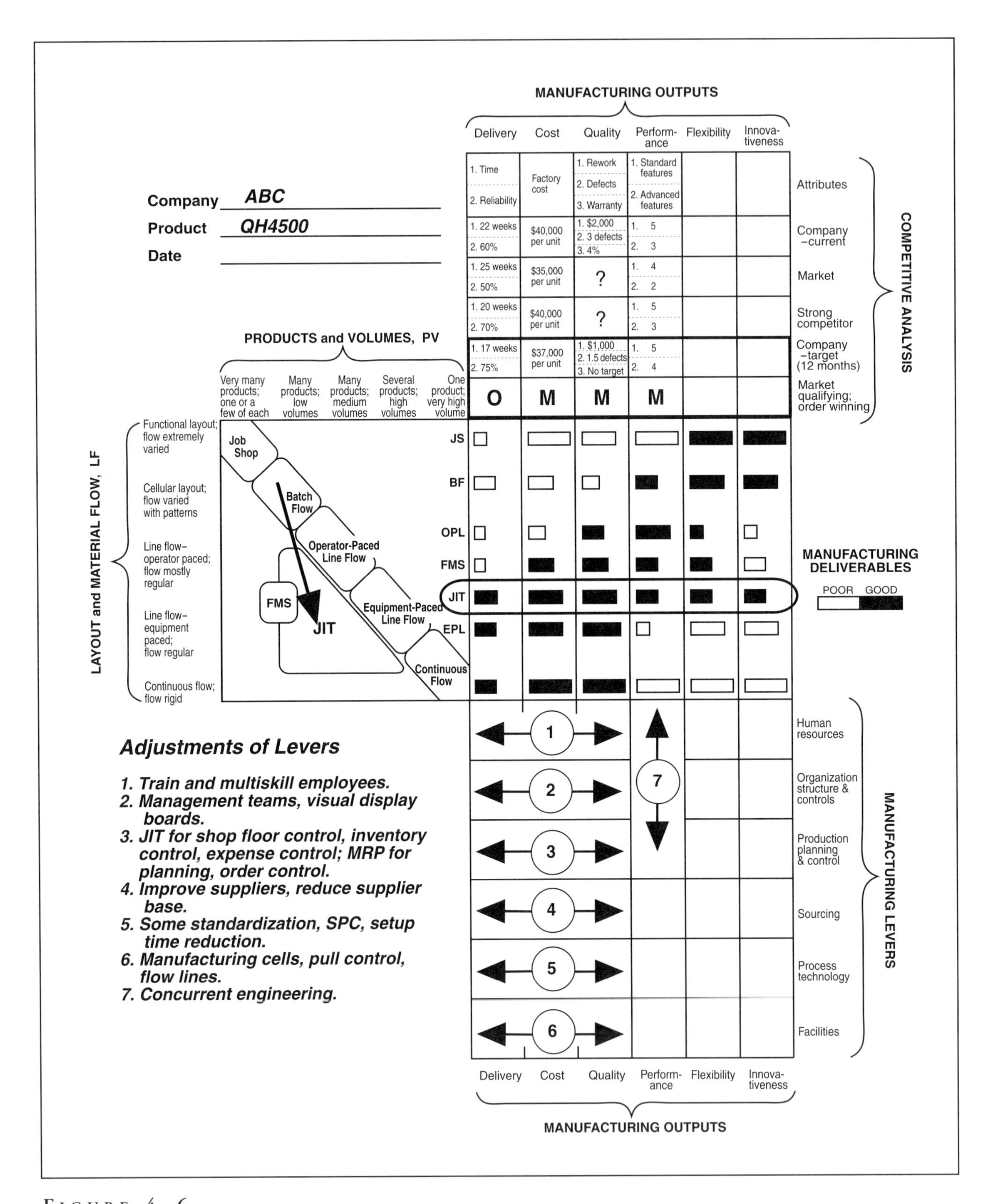

FIGURE 4–6

Adjusting the Levers for the New Production System

SUMMARY

Manufacturing provides six outputs—cost, quality, performance, delivery, flexibility, and innovativeness—to its customers. The levels at which the outputs are provided depend primarily on the production system used. No production system can provide the highest levels of all six outputs. The outputs that manufacturing chooses to provide must be those that its customers want.

Competitive analysis is the element of manufacturing strategy wherein a determination is made of what manufacturing outputs the customer wants and what target levels should be set. A simple and effective way to perform a competitive analysis is outlined in this chapter. It requires the following types of data:

- The organization's products
- Competitors' products
- Customer needs and expectations
- The organization's current production system

Outcomes from the competitive analysis element of manufacturing strategy are 1) the market qualifying and order winning manufacturing outputs, with their target levels, for each product or product family, and 2) a production system that can provide the outputs at the target levels and can be achieved by the organization. Several attempts or iterations are usually required to complete the competitive analysis and find a suitable set of outputs and production system.

FURTHER READING

Fine, C., and A. Hax, "Manufacturing Strategy: A Methodology and an Illustration," *Interfaces*, Vol. 15, No. 6, pp. 28–46, 1985.

Hill, T., *Manufacturing Strategy: Text and Cases*, Homewood, IL: Richard D. Irwin, 1989.

Skinner, W., "Manufacturing—Missing Link in Corporate Strategy," *Harvard Business Review*, May–June 1969.

Appendix: Another Instrument for Doing a Competitive Analysis

Simpler frameworks than that shown in Figure 4-4 can be used to do a competitive analysis. One such framework, shown in Figure 4-7, was reported by Fine and Hax (1985) for use in a wire and cable business unit at Packard Electric. The framework assesses the relative importance and current level of performance of four manufacturing outputs—cost, quality, delivery, and flexibility—for three product families. The relative importance of each manufacturing output is assessed by assigning a weight to it so that the weights for all four outputs add up to 100 percent. An assessment is then made of the level at which each output is provided relative to the organization's competitors.

This framework is a simpler version of the one used in this book (Figure 4-4). Those manufacturing outputs with the highest weights may be called order winning outputs. The others may be market qualifying outputs. This would mean that quality is the order winning output and cost and delivery are market qualifying outputs for each of the three product families. Flexibility is relatively unimportant. The level at which each output is currently provided would be considered, with the importance of the output, to determine the target levels for each output for each product family.

Manufacturing Unit *Wire and Cable Business Unit, Packard Bell*

Product Family	COST		QUALITY		DELIVERY		FLEXIBILITY	
	Importance	Current Level	Importance	Current Level	Importance	Current Level	Importance	Current Level
1. Cable	30	Very strong	40	Weak	20	Even	10	Weak
2. Printed circuits	20	Very weak	50	Even	20	Strong	10	Strong
3. Copper rod	20	Strong	40	Strong	30	Very strong	10	Weak

Source: Adapted from Fine and Hax 1985.

Figure 4–7

A Simple Tool for Competitive Analysis

CHAPTER 5

LEVEL OF MANUFACTURING CAPABILITY

Changes are adjustments to manufacturing levers. Some companies have no difficulty making changes, even very large ones. Others struggle to make even a small change. Two factors have a major effect on manufacturing's ability to make changes: top management commitment and level of manufacturing capability.

Top Management

Top management commitment and active participation are required if significant changes are to be made. Harley-Davidson had both in the 1980s when it turned itself around. In 1981, AMF sold Harley-Davidson to a group of Harley-Davidson managers. High interest rates, a recession, strong competition from Japanese motorcycle manufacturers, and quality problems contributed to a bleak outlook for Harley-Davidson's future. Yet by 1989, the company had rebounded. Its share of the over 1,000 cc motorcycle market had climbed from a low of 16 percent up to 25 percent. The levels of the manufacturing outputs were up: costs were down, quality was way up, and delivery time was much improved. Harley-Davidson made dramatic improvements in its manufacturing capabilities when it implemented a broad range of JIT and quality control techniques. These changes would have been impossible without the commitment and active participation of top management, which existed at Harley-Davidson because of the extreme difficulties facing the company.

Manufacturing Capability

New manufacturing capabilities are built on existing manufacturing capabilities. The smaller this base is, the more difficult it is to build on. When the level of manufacturing capability is low, anything beyond a few small changes is difficult. As the level of capability increases, more changes can be made at a faster pace.

The IBM RTP facility located in North Carolina provides an example of how a solid base of manufacturing capability makes it easier to implement new techniques and technologies. The RTP facility uses almost every advanced manufacturing technique and technology, including just-in-time, concurrent engineering, a flexible workforce, employment security, training, robots, automatic guided vehicles, flexible manufacturing systems, continuous flow production, product standardization, and so on. It is easy for RTP to make changes and add new techniques to this solid base. This high level of capability enables RTP to provide high levels of the manufacturing outputs.

Measuring the Overall Level of Manufacturing Capability

A high level of manufacturing capability is important for two reasons:

- It enables a production system to provide higher levels of the manufacturing outputs.
- It permits changes to be made quickly and easily.

A measure of the overall level of manufacturing capability is shown in Figure 5-1. The measure can take any value from 1.0 to 4.0. A value of 1.0 indicates an infant level of capability, 2.0 indicates an industry average level, 3.0 is an adult level, and 4.0 is a world class level of manufacturing capability. Harley-Davidson's overall manufacturing capability is near the adult level. IBM RTP is near the world class level. Organizations with manufacturing capabilities near adult and world class levels can implement a great deal of change quickly. Organizations with infant and

Infant *Level 1*	Average *Level 2*	Adult *Level 3*	World Class *Level 4*
The production system makes little contribution to the organization's success.	Manufacturing is satisfied to keep up with its competitors and maintain the status quo.	The production system provides market qualifying and order winning outputs at target levels.	The production system strives to be the best in the world in all activities in all manufacturing subsystems (levers).
Manufacturing is low tech and unskilled.	Manufacturing consists of standard, routine activities.	All manufacturing decisions are consistent with the manufacturing strategy.	The production system is a major source of competitive advantage.

FIGURE 5–1

The Overall Level of Manufacturing Capability

industry average levels of capability can make only small changes at a slow rate.

The concept of a continuum of manufacturing capability is adapted from a similar concept introduced by Hayes and Wheelwright in their 1985 article, "Competing Through Manufacturing," where they describe a range of strategic roles that manufacturing can play in an organization. (See the Further Readings section at the end of the chapter.)

AN INFANT OVERALL LEVEL OF MANUFACTURING CAPABILITY

Many production systems begin with an infant level of manufacturing capability. For example, an organization may be formed to sell a new product or take advantage of a market niche and so is strongly oriented toward marketing or product design. Manufacturing is simply a necessary evil. Products are produced in general-purpose facilities on a poorly managed production system using as many purchased parts and materials as possible. Manufacturing personnel are unskilled and unsophisticated. As these organizations mature, they notice that their competitors manufacture in a way that results in lower cost, higher quality, and better delivery, performance, flexibility, and innovativeness. They then make improvements to their production systems to raise their level of manufacturing capability so that higher levels of the outputs can be provided.

An Industry Average Overall Level of Manufacturing Capability

In a production system with an industry average level of manufacturing capability, manufacturing looks and behaves like its competitors. The organization and its competitors have similar facilities, employ similar process technologies (purchased from the same vendors), use the same suppliers, follow industrywide employment practices, and so on. Economies of scale are pursued actively. New techniques and technologies are adopted only when everyone in the industry is using them.

An industry average level of manufacturing capability is sufficient when there is a stable set of competitors and a growing market. It becomes insufficient when new competitors appear or the market stops growing and there is excess capacity in the industry. Sometimes offshore competitors enter the market and gain market share when they provide higher levels of the manufacturing outputs (as a consequence of their higher manufacturing capabilities or different production systems). Organizations with industry average levels of capability respond by making improvements to their production system to increase their capability.

An Adult Overall Level of Manufacturing Capability

When manufacturing capability is at an adult level, the analysis and planning principles outlined in this book are applied. Plants-within-plants are organized, each of which has a production system capable of providing the required levels of the market qualifying and order winning outputs, as determined by a competitive analysis. A long-term perspective is adopted. Achieving an adult level of manufacturing capability is quite an accomplishment, but some production systems strive for even higher levels of capability.

A World Class Overall Level of Manufacturing Capability

A production system with a world class level of manufacturing capability can provide more than just one manufacturing ouput at an order winning level. It may, for example, seek to be the highest quality, lowest cost producer of products in the world.

Providing two or more outputs at the highest levels in the world gives an organization a major competitive advantage over other manufacturers who, with a lower level of capability, have difficulty providing even one output at an order winning level.

Production systems with world class capabilities develop much of their own process technology because their requirements and expertise exceed the capabilities of equipment vendors. They seek to acquire expertise in new technologies before the technologies are proven. Like IBM RTP, these organizations invest equally in soft technologies (such as training, changing organizational structures, improving operating procedures, changing layouts, and so on) and hard technologies (such as new facilities and equipment). See Chapter 10.

Level of Manufacturing Capability for Each Manufacturing Lever

The overall level of manufacturing capability for a production system is the sum of the manufacturing capabilities of each manufacturing subsystem, or lever—human resources, organization structure and controls, sourcing, production planning and control, process technology, and facilities. The higher the manufacturing capability of each lever, the higher the overall capability of the production system. The manufacturing capability of a lever is also measured on a four-point scale: 1—infant, 2—industry average, 3—adult, 4—world class. Exactly what constitutes each level of each lever will vary from industry to industry. A tool like benchmarking is used to make these determinations (see Chapter 9). The scale shown in Figure 5-2 is a starting point for measuring the level of capability for a lever. It is modified for use in a particular company as benchmarking data is collected.

The level of manufacturing capability need not be the same for each lever. For example, an organization may have industry average facilities and process technology, an adult level of capability in human resources, and an infant level of capability in production planning and control. When some levers in a production system have lower levels of capability than others, the overall level of capability is diminished. Manufacturing strategy identifies these levers and the adjustments (that is, the changes and improvements) needed to raise the lower levels of capability. In most cases, the goal

	Level of Manufacturing Capability			
Manufacturing Levers	**Infant** *Level 1*	**Average** *Level 2*	**Adult** *Level 1*	**World Class Level 4**
Human resources		• Employees are an expense • Unskilled • Human robots		• Employees are an investment • Multiskilled • Problem identification and solving
Organization structure and controls		• Hierarchical, centralized • Cost accounting driven performance measures • Staff is very important		• Flat, decentralized • Competitive performance measures • Line is very important
Production planning and control		• Centralized, complex • Detailed monitoring of resource usage		• Decentralized, simple • Aggregate monitoring of resource usage
Sourcing		• Large number of suppliers • Short-term contracts • Lowest cost		• Small number of suppliers • Partnership, full responsibility • Critical capabilities
Process technology		• Mature technology • Developed externally • Reduce cost		• Modern soft and hard technologies • Developed internally • Provide manufacturing outputs
Facilities		• General purpose • Large, infrequent changes • Capital appropriation driven		• Focused • Frequent, incremental changes • Improve capabilities

FIGURE 5–2

Levels of Manufacturing Capability for Each Manufacturing Lever

is to have a production system where all the levers have the same high level of capability.

An Industry Average Level of Capability for a Manufacturing Lever

A production system with an industry average level of manufacturing capability tends to have a relatively unskilled and closely supervised workforce that is not involved in decision making or problem solving and has little employment security. The organizational structure is hierarchical, decision making is centralized, and staff groups are very influential. Performance measurement systems are based on cost accounting measures.

Production planning and control systems are centralized and include complex procedures for tracking the uses of materials and other resources. A large network of suppliers is used, and multiple sources are maintained for each purchased item. Suppliers are kept at arm's length. Short-term contracts are awarded on the basis of lowest cost. Process technology is mature. The production system's attitude toward technology is that technology is developed by equipment vendors and is used to reduce costs. Equipment and facilities are general purpose. Changes occur infrequently; when they do occur, they are large. The rate of change is controlled through a rigid capital appropriation request procedure.

A World Class Level of Capability for a Manufacturing Lever

A production system with world class manufacturing capability is quite different. Human resources are seen as an investment, an asset to be improved through training. Human resources policies include employment security and involvement of the workforce in problem solving. Decision making is pushed to the lowest level of competence (that is, the lowest level in the organization where employees have the competence to make effective decisions). Training is used to raise the level of competence in all parts of the production system. The organizational structure is flat, line groups are more influential, and performance measures are tied to providing high levels of the market qualifying and order winning outputs.

Production planning and control systems are decentralized and simple. Just-in-time techniques such as pull control systems and setup time reduction are used. Because delivery times are short, procedures for monitoring and controlling shop floor activities are limited to aggregate control of resources released to production and monitoring of products produced. Partnerships with a small network of suppliers possessing critical capabilities are developed. Long-term contracts are the norm. Suppliers are given responsibility for improving quality, reducing cost, improving product design, and so on. Process technology is often developed

SITUATION 5.1

The Honda-Yamaha War (Continued)

A HIGHER LEVEL of manufacturing capability was one of the reasons that Honda was able to provide higher levels of the market qualifying and order winning outputs than Yamaha. Yamaha had just completed a period of expansion during which new facilities, processes, employees, suppliers, and systems were started up. One consequence of the expansion was that the existing manufacturing capability was spread over a larger number of sites and operations. This dilution of expertise resulted in a reduction in Yamaha's overall level of manufacturing capability.

Manufacturing capability profiles for Honda and Yamaha at the time of the Honda-Yamaha war are shown in Figure 5-3. The level of manufacturing capability for each of the first three levers was 3.5 for Honda and 2.5 for Yamaha. The lower figures for Yamaha were a consequence of its expansion. With respect to the level of manufacturing capability of suppliers, Honda's suppliers were the best in the industry. The level of manufacturing capability for process technology and facilities was high for Yamaha because many of the processes and facilities were new. Although the processes and facilities at Honda were older, many improvements had been made to them through Honda's improvement programs, and so the level of manufacturing capability was also high.

The manufacturing capability profile for Honda was better than that for Yamaha. In addition, those levers that most affect the flexibility and innovativeness outputs—human resources, organization structure and controls, sourcing, and production planning and control—had higher levels of capability at Honda than at Yamaha. The higher level of manufacturing capability and production systems that were better able to provide the outputs customers wanted were the reasons why Honda won the Honda-Yamaha war.

internally. Technology is seen as a means of providing higher levels of the market qualifying and order winning outputs. Soft

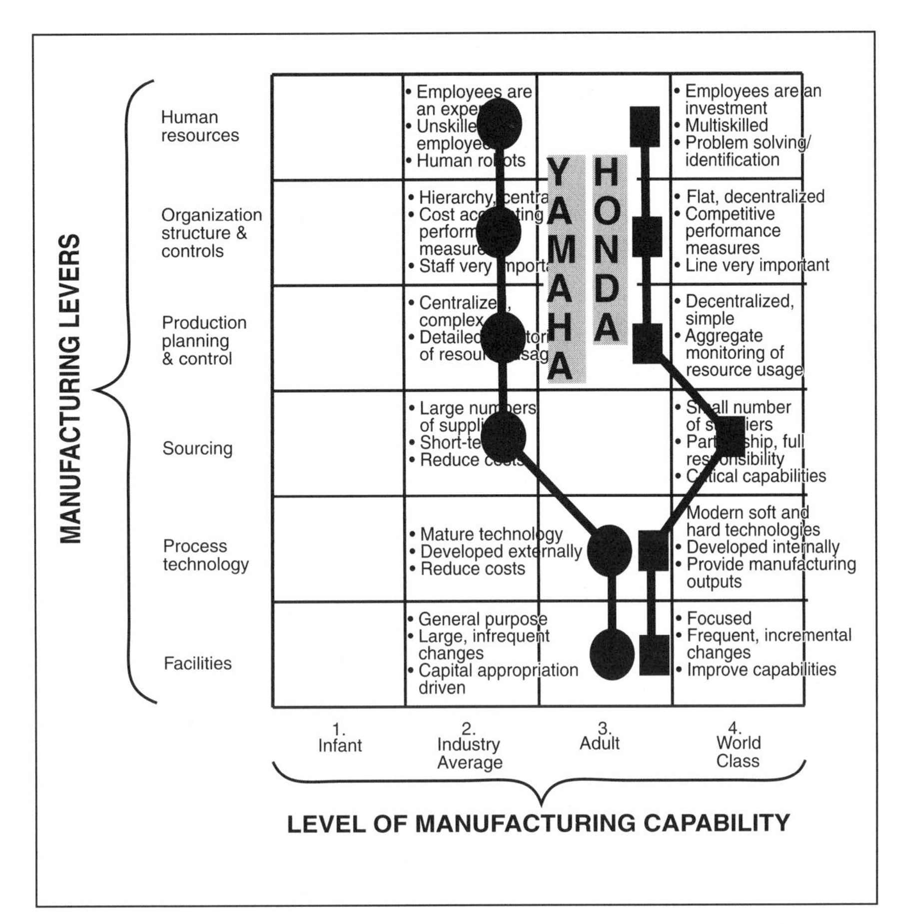

FIGURE 5–3

Profiles of Manufacturing Capability for Honda and Yamaha during the Honda-Yamaha War

and hard technologies are equally important (see Chapter 10). Facilities and equipment are more specialized. Frequent, incremental improvements are made to raise the level of manufacturing capability so that higher levels of the outputs can be provided.

A quick test to assess whether a production system has achieved a world class level of manufacturing capability consists of the following two checks:

1. *How is new process technology developed?* A production system with a world-class level of capability will be actively involved (alone or in partnership with others) in developing new process technology. The technology cannot be bought from equipment vendors because the production system's capability and requirements are much higher than those of the vendors. The production system is also interested in acquiring expertise in new technologies before the technologies are proven.

2. *How are employees compensated?* The compensation systems used in production systems with world class levels of capability have the following characteristics. Incentives (such as gainsharing and profit sharing) are an important part of the compensation package and are used to encourage teamwork. Compensation is linked to mastery of skills. All employee groups sacrifice equally during economic downturns. (In contrast, the compensation systems used in production systems with lower levels of capability are characterized by incentives based on individual output, pay scales based on job classifications, and proportionately more layoffs of hourly employees compared to staff employees during economic downturns.)

Assessing the level of manufacturing capability is the final element of the manufacturing strategy. Its place among the other elements is shown in Figure 5-4. This manufacturing strategy worksheet can now be used to analyze manufacturing and to develop strategies for improving it.

Manufacturing's first job is to provide the market qualifying outputs at the target levels. The second job is to take a market qualifying output and provide it at a much higher level, the order winning level. Some organizations are not content to provide just one output at an order winning level. They want two or more order winning outputs. This can be achieved only when the organization's production system has a world class level of manufacturing capability. Toyota is an example of such an organization. It is able to provide two order winning outputs, quality and performance, while still keeping other outputs, such as cost, at market qualifying levels. Two Toyota marketing slogans, "There's quality and there's Toyota quality" and "Toyota, I love what you do for me," suggest that Toyota is trying to provide quality and performance at the highest possible levels. Toyota can do this as a result of more than 20 years of work developing a new production system called JIT and raising the levels of manufacturing capability of each lever in the JIT production system to world class levels.

Assessing the level of manufacturing capability requires information about practices and processes in the organization, at other

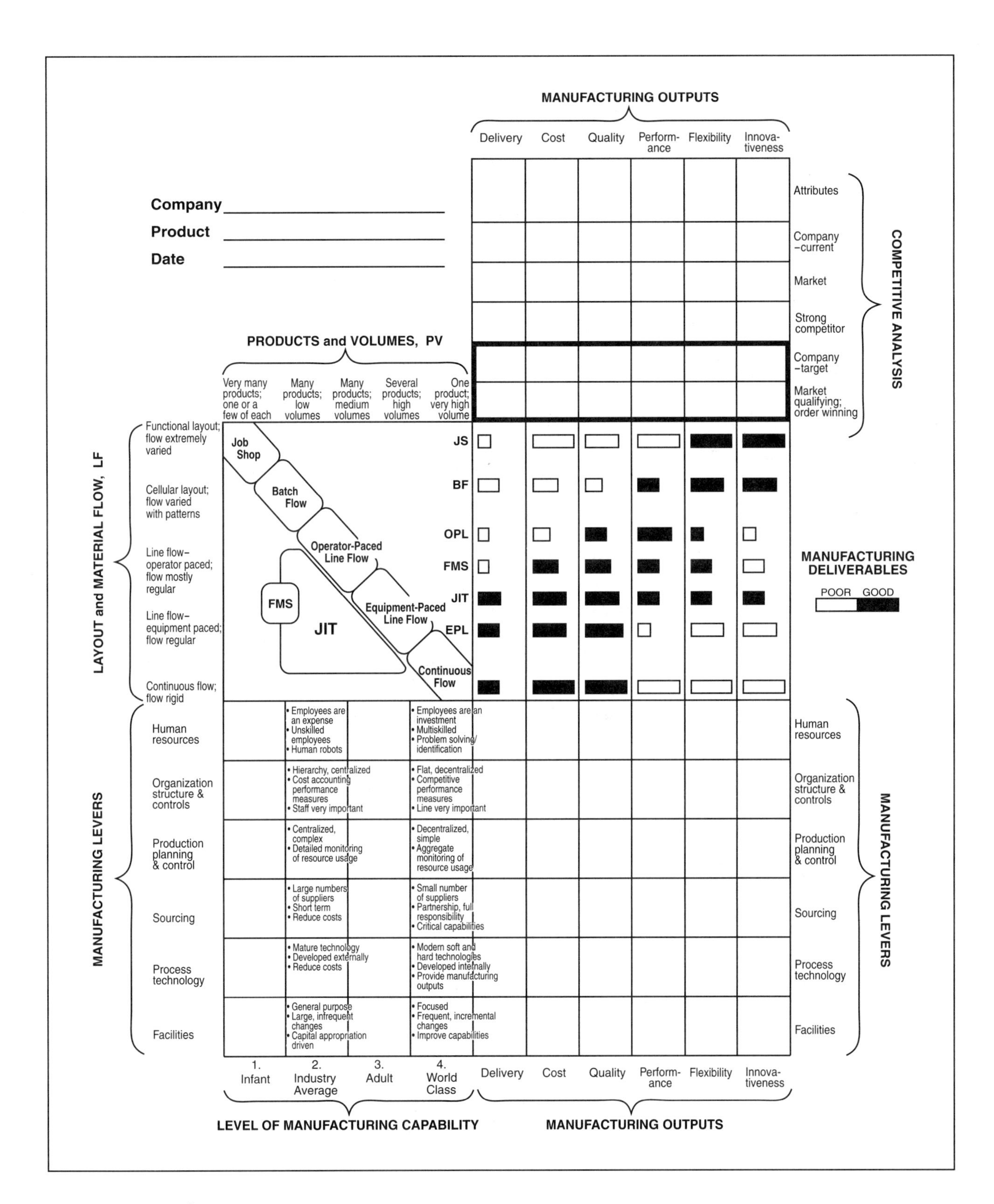

FIGURE 5–4

Adding the Level of Manufacturing Capability to the Strategy Worksheet

SITUATION 5.2

DEC Raises Its Level of Manufacturing Capability[1]

DESPITE a manufacturing function whose capability was at an infant level compared to that of other industries, Digital Equipment Corporation (DEC) enjoyed great success with its technologically innovative products.This changed in 1983. Shipments were missed, customer service slipped, and stock prices fell. Digital responded by raising its level of manufacturing capability and transforming manufacturing into a powerful competitive weapon. The company defined its own levels of manufacturing capability.

Level 1: An Infant Level of Capability

In 1983, Digital's capability was at the infant level. This prevented the company from providing the required outputs—cost, quality, performance, delivery, flexibility, and innovativeness—at market qualifying levels.

Level 2: Class "A" MRP II

Digital wanted manufacturing to do things well. It set a goal to achieve Class "A" MRP II certification. This would force improvements to be made, thereby ensuring that all manufacturing activities were done properly. By 1986, 25 Digital facilities had achieved this certification.

Level 3: TQC/JIT

The level of capability continued to rise through the late 1980s. Improvement efforts moved beyond bill of material and inventory accuracy to quality and cycle time. A total quality commitment (TQC) approach with JIT techniques was incorporated into the production systems so that the highest possible levels of the market qualifying and order winning outputs could be provided.

Level 4: World Class Level of Capability

By 1989, Digital was beginning to move beyond level 3 to a world class level of manufacturing capability. Although a definition of world class for Digital had yet to be formulated, some of the desired characteristics had been identified. They included benchmarking, continuous improvement, and employee involvement.

The speed with which improvements were accomplished was impressive. Figure 5-5 shows some results for a plant in Burlington, Vermont.

Performance Measure	1985	1986	1987	1988	1989
Manufacturing cycle time	10 weeks	2 weeks	5 days	3 days	2.5 days
Master schedule performance	75%	85%	97%	97%	daily
WIP inventory turns	less than 20	less than 20	23	49	65

FIGURE 5–5

Some Results from a DEC Plant

organizations in the same industry, and at organizations in other industries. Manufacturers who do not look outside their own organization tend to overrate their own capabilities. Benchmarking is a tool that can help assess the level of capability of each manufacturing lever. It is also used to discover new practices and processes that can raise the level of capability. Benchmarking and improvement approaches that help to raise the level of capability, such as total quality management, short cycle manufacturing, and kaizen, are discussed in Chapter 9.

Strategy can be thought of as a process of matching internal capabilities with external opportunities. Failing to achieve a match, by over- or underestimating manufacturing capability or misreading the environment, can have devastating consequences. A dramatic illustration of what can go wrong is the case of Rolls-Royce's aircraft engine division. They overestimated their manufacturing capability so badly in the late 1960s that by 1971 they were bankrupt.

Situation 5.3

Rolls-Royce Overestimates Its Level of Manufacturing Capability[2]

ROLLS-ROYCE was once the most famous engineering firm in the world. Although known to the general public as a producer of luxury automobiles, its major product was aircraft engines. For decades, Rolls-Royce had performed magnificently in this field. Among other accomplishments, it produced the Merlin piston engine, which powered the Spitfires and Hurricane fighters of World War II, as well as the first jet engine used in commercial aircraft.

In the mid-1960s Rolls-Royce found itself in a dilemma. Its major aircraft engines were in the mature and declining stages of their product life cycles (see Chapter 11) and the company had no new products to take their places. Several new engine programs had been initiated, but these were for limited markets. Another engine manufacturer had been acquired, but that firm's capabilities and order book did not solve many problems for the core Rolls-Royce operation.

In this gloomy situation, the U.S. civil aviation market looked like a saving opportunity. The new wide-body Douglas and Lockheed aircraft were in early development, and Rolls-Royce was determined to enter this market with a new engine. The bidding was intense. Customers were much more cost- and delivery-conscious than Rolls-Royce's traditional military and government customers, and Rolls-Royce was competing against established firms such as Pratt and Whitney and General Electric. By 1968, Rolls-Royce succeeded in obtaining a major order from Lockheed, but at a low fixed price with significant late delivery penalties.

The euphoria of the sale dissipated quickly in subsequent months as Rolls-Royce began to realize that it did not have the design and manufacturing capabilities required for the project. Major difficulties soon appeared in product and process technology, organization structure and controls, and production planning and control.

Product and Process Technology

The new engine incorporated new, unproven technical advances. This was difficult for Rolls-Royce, which, throughout its history, had been a technology follower rather than a technology leader. Unanticipated problems and delays were encountered as both internal and supplier capabilities were overburdened. Development costs, which had originally been estimated at £65 million, doubled to £135 million by early 1970 and nearly doubled again in 1971 to £220 million.

Organization Structure and Controls

Rolls-Royce was a company dominated by engineers. Financial issues were often subordinated to technical issues. A lack of financial sophistication and the urgency of the project permitted serious errors in estimating and allowed the full consequences to go unnoticed.

Production Planning and Control

While Rolls-Royce was well equipped to handle complex, ongoing production tasks, it was not equipped to handle complex development projects.

In 1971, Rolls-Royce fell into bankruptcy, mostly as a consequence of having badly overestimated its capabilities.

SUMMARY

The levels at which the manufacturing outputs—cost, quality, performance, delivery, flexibility, and innovativeness—are provided depend on the production system used and the level of that production system's manufacturing capability. A production system's level of capability is the sum of the levels of capability of each manufacturing subsystem or lever. The levels of capability of each lever in the production system need not be the same.

Manufacturing capability is measured on a continuous scale ranging from 1.0 to 4.0: 1.0 is an infant level of capability, 2.0 is an industry average level, 3.0 is an adult level, and 4.0 is a world class level. Manufacturing capability is the foundation on which changes and improvements are built. The higher the level of manufacturing capability, the easier it is to make changes and improvements. Organizations with a world class level of capability can provide more than one manufacturing output at an order winning level. This gives them a significant advantage over competitors who, hampered by low levels of capability in many levers, struggle to provide even one output at an order winning level.

NOTES

1. Adapted from an article by P. E. Moody. See the Further Reading section below.
2. See *The Financial Times*, London, England, August 3, 1973, p. 16.

FURTHER READING

Fine, C., and A. Hax, "Manufacturing Strategy: A Methodology and an Illustration," *Interfaces*, Vol. 15, No. 6, pp. 28–46, 1985.

Hayes, R., and S. Wheelwright, "Competing Through Manufacturing," *Harvard Business Review*, pp. 99–109, January–February 1985.

Moody, P. E., "Digital Equipment Corporation: Journeying to Manufacturing Excellence," pp. 175–186 in *Strategic Manufacturing*, P. E. Moody (ed.), Homewood, IL: Dow-Jones Irwin, 1990.

APPENDIX: ANOTHER INSTRUMENT FOR ASSESSING MANUFACTURING CAPABILITY

Many instruments are available for helping an organization assess its level of manufacturing capability. A different instrument from that presented in this chapter is shown in Figure 5-6. It is adapted from an instrument proposed by Fine and Hax (1985) for use in a wire and cable business unit at Packard Bell. The instrument summarizes the current policies, strengths, and weaknesses of each manufacturing lever in the production system. Two limitations of this instrument compared to the one developed in this chapter are the following: 1) There is no relative assessment of the organization's policies against those of its competitors, and 2) it is not easy to relate these policies to the other elements of manufacturing strategy.

Manufacturing Lever	Description of Past Policy	Strengths	Weaknesses
Human resources	Strong quality of work life programs	Employee participation in decisions Good communication	Compensation system does not consider quality
Organization structure and controls	Control systems have short-term, tactical orientation Respond to GM in principal lines	Good control orientation Low risk	Shortsighted system Reactive rather than anticipatory, focus concept ignored
Sourcing	Significant backward integration — all the way to wire rod	Good control over cost and quality	Less focus, transfer pricing complications
Production planning and control	Use overtime, third shift, and inventory to respond to cyclical demand	Flexibility	Layoffs and overtime are costly
Process technology	Cable and copper in automated continuous process Printed circuits in job shop Heavy use of statistical process control and cost of quality tools	State-of-art in cable and copper Integrated approach, top management support	Automation in printed circuits could reduce costs Quality lags relative to Japanese competition
Facilities	Process focus	Economies of scale	Long physical supply distances

Source: Adapted from Fine and Hax 1985.

FIGURE 5–6

A Tool for Assessing Existing Manufacturing Policies

CHAPTER 6

The Complete Framework for Formulating Manufacturing Strategy

A complete framework is now available for analyzing manufacturing and developing a strategy for improving it. The worksheet for this framework is shown again in Figure 6-1. It can be used in many ways for many purposes, including:

- Analyzing an existing operation
- Generating and evaluating alternate strategies
- Analyzing competitors' strategies
- Developing a complete manufacturing strategy

One procedure for developing a complete manufacturing strategy, which has worked well in practice, consists of the following three steps.

Step 1: Where am I?

- Determine manufacturing's current location on the PV–LF matrix, and the *production system* in use.
- Assess the current level of capability for each manufacturing lever using the *manufacturing capability* section of the worksheet.

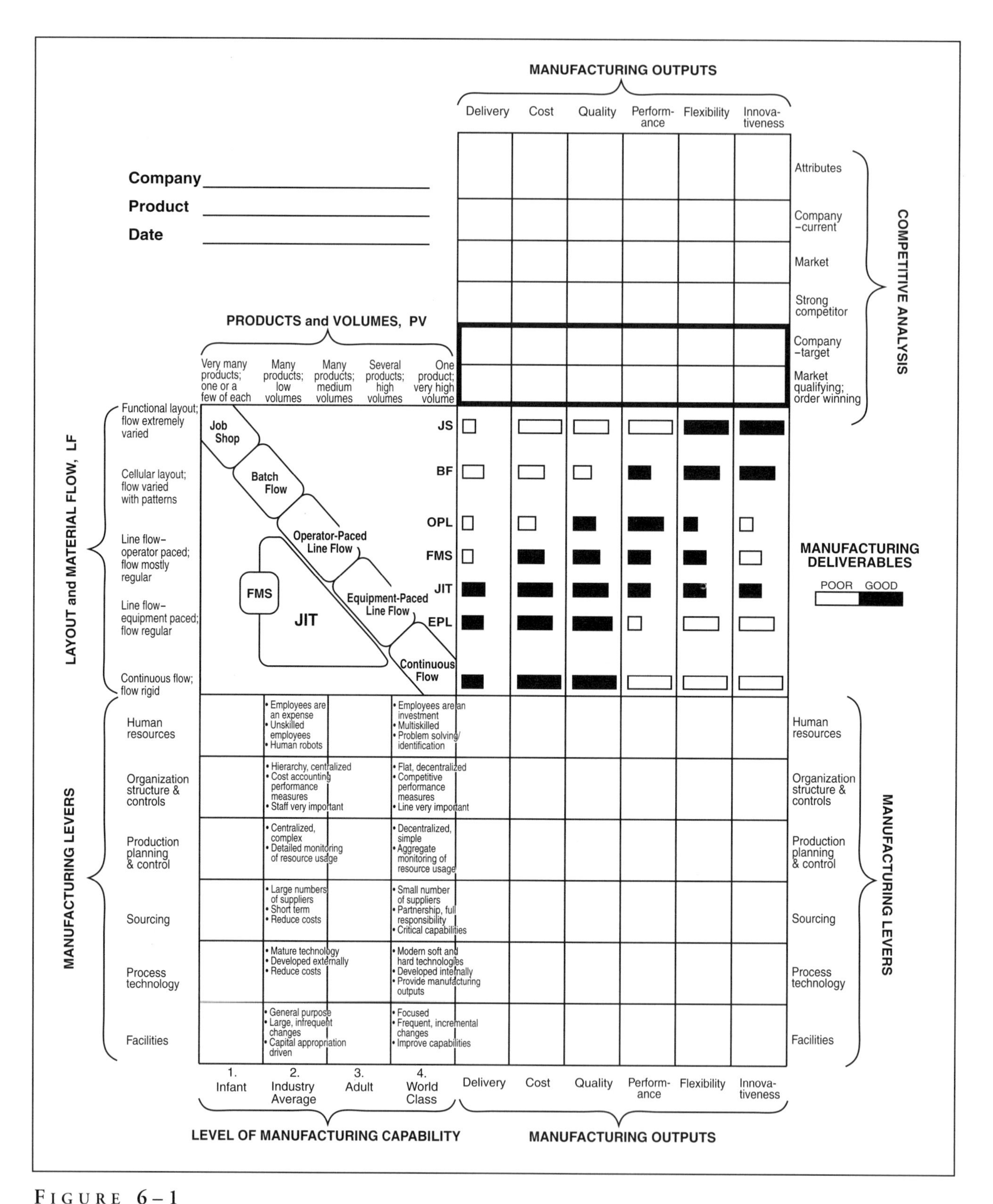

FIGURE 6–1

The Manufacturing Strategy Worksheet

Step 2: Where do I want to be?

- Complete a *competitive analysis* to determine the market qualifying and order winning outputs that must be provided by the production system, and set 12-month target levels for them.
- Find the row of outputs on the *manufacturing deliverables* chart that best matches the required market qualifying and order winning outputs.
- Determine the *production system* on the PV–LF matrix that best provides these manufacturing outputs.

Step 3: How will I get from where I am to where I want to be?

- If the production systems determined in steps 1 and 2 are the same, then adjust the *manufacturing levers* on the levers section of the worksheet so that the production system is better able to provide the market qualifying and order winning outputs at the target levels. Make sure that these adjustments are possible with the current level of manufacturing capability.
- If the production systems determined in steps 1 and 2 are not the same, then make adjustments to the *manufacturing levers* on the levers section of the worksheet, so that:
  - ➤ The current production system changes to the desired production system
  - ➤ The required market qualifying and order winning outputs are provided at the target levels
  - ➤ The adjustments can be made with the current level of manufacturing capability.

  If this cannot be done, then return to step 2. Select different market qualifying and order winning outputs and repeat step 3.

To illustrate how this procedure works, we consider again the ABC plant from Chapter 4.

Using the Complete Framework at ABC Company

The problem at the ABC plant was an existing batch flow production system that was not able to provide the new market qualifying and order winning outputs at the target levels.

Step 1: Where is the ABC plant?

The current production system at the ABC plant was a batch flow system (see Figure 6-2). It provided high levels of flexibility and innovativeness—exactly what was required in the past when four families of custom-engineered products were produced by one production system. The current level of manufacturing capability for each lever in the batch flow production system is also shown in the figure. Four of the levers—organization structure and controls, production planning and control, sourcing, and process technology—had industry average levels of manufacturing capability. The human resources lever was slightly better than industry average because the plant was located in a rural area and the employees had a good work ethic. Because the facilities were new and the equipment was modern, the facilities lever had an adult level of manufacturing capability. The overall level of manufacturing capability for the ABC plant was estimated to be 2.2, slightly better than industry average. An important consequence of this modest level of capability was that the plant would be able to implement only a small number of changes at any one time. To summarize, the results after step 1 were:

Current production system—Batch flow

Current manufacturing capability—Slightly better than industry average

Step 2: Where does the ABC plant want to be?

The competitive analysis for product family QH4500 was discussed in Chapter 4. The market qualifying outputs were cost, quality, and performance. Delivery was the order winning output. Twelve-month target levels for these outputs were determined and are shown again in Figure 6-3. The area of the manufacturing deliverables chart where these outputs are provided most easily is shown in the figure. The area corresponds to the FMS, equipment-paced line flow, and JIT production systems. Any one

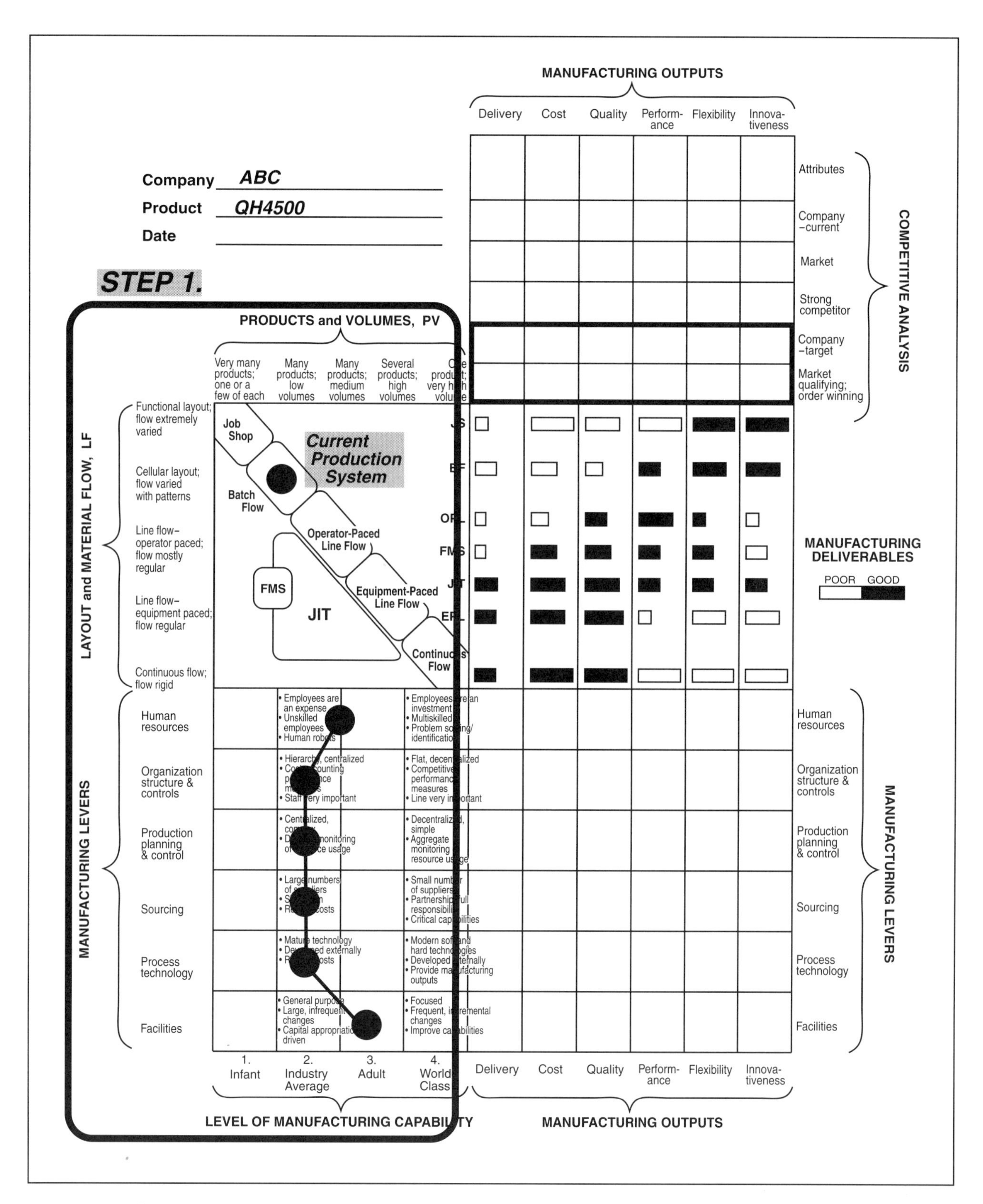

FIGURE 6–2

Step 1. Where Is the ABC Plant?

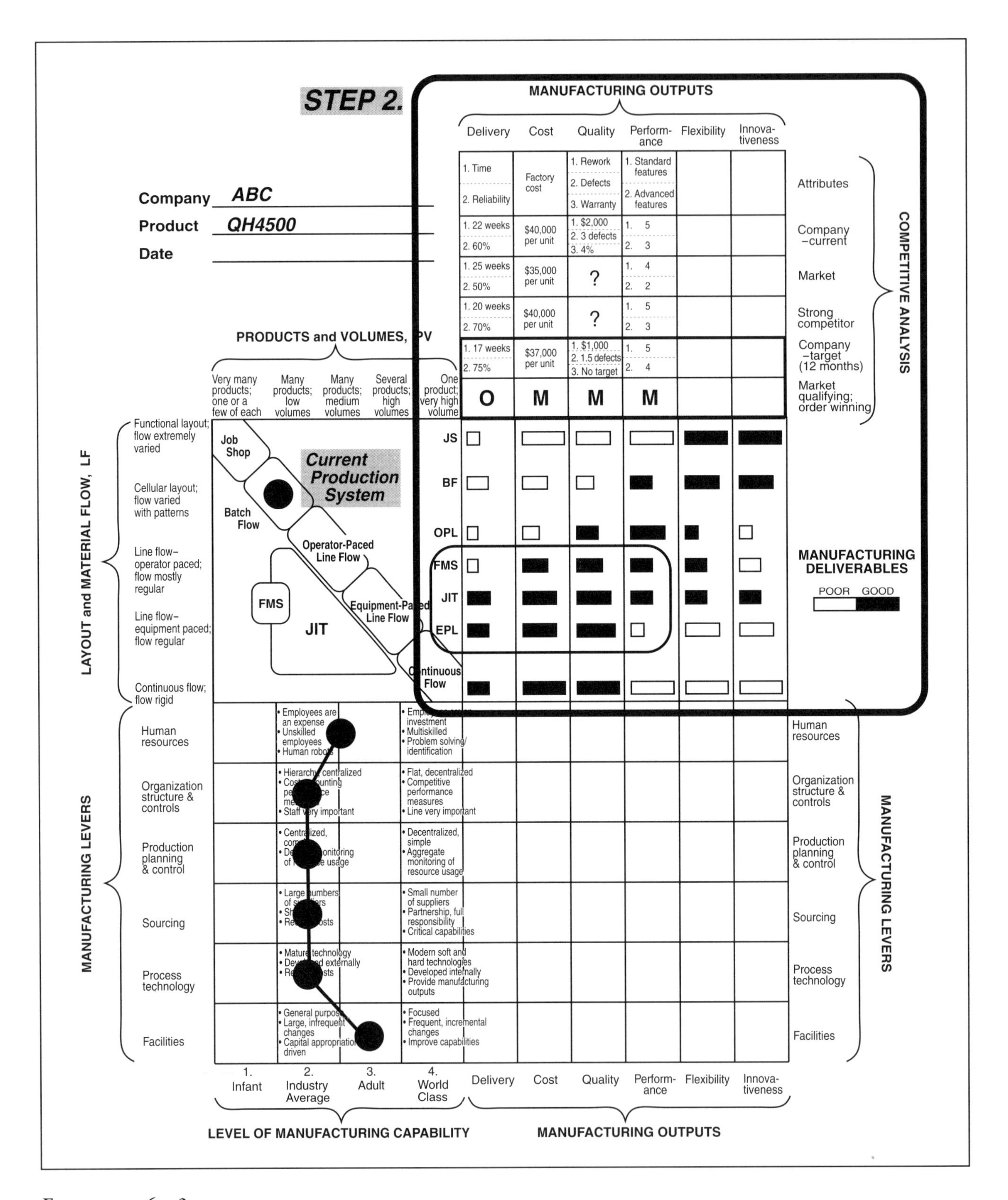

	Delivery	Cost	Quality	Performance	Flexibility	Innovativeness
Attributes	1. Time 2. Reliability	Factory cost	1. Rework 2. Defects 3. Warranty	1. Standard features 2. Advanced features		
Company –current	1. 22 weeks 2. 60%	$40,000 per unit	1. $2,000 2. 3 defects 3. 4%	1. 5 2. 3		
Market	1. 25 weeks 2. 50%	$35,000 per unit	?	1. 4 2. 2		
Strong competitor	1. 20 weeks 2. 70%	$40,000 per unit	?	1. 5 2. 3		
Company –target (12 months)	1. 17 weeks 2. 75%	$37,000 per unit	1. $1,000 2. 1.5 defects 3. No target	1. 5 2. 4		
Market qualifying; order winning	O	M	M	M		

FIGURE 6–3

Step 2. Where Does the ABC Plant Want to Be?

of these would provide the required outputs for QH4500. Overlooking product and volume considerations for the moment, the best choice would be an equipment-paced line flow production system. This is a well-known production system and it is relatively easy to design, install, operate, and manage. JIT and FMS are not nearly as appealing. JIT is a very difficult production system to design, operate, and manage (see Chapter 17), and an FMS requires a large capital investment and highly skilled employees (see Chapter 15). The QH4500 product family consisted of many custom engineered products produced in low volumes, so it could not be produced on an equipment-paced line flow production system. (See the PV–LF matrix in Figure 6-3.) The ABC plant tried to change the number of products and the production volumes for QH4500 so that it could be produced on an equipment-paced line flow system.

- The plant asked the product design group to standardize and modularize QH4500. These two changes would reduce the number of different products that had to be produced. The design group refused, further straining an already poor relationship between design and manufacturing.
- The plant asked the corporation to give it a larger share of the international market. This would increase production volume. The request was refused because of the plant's history of losing money.

The message to the plant seemed to be: "Before manufacturing can ask other parts of the organization to make changes, it will have to make its own changes and achieve significant improvements. Once it does that, it will earn the right to ask other parts of the organization to change."

Because the ABC plant would have to produce many products in low to medium volumes, it would have to use an FMS or a JIT line flow production system. FMS was ruled out because the plant could not afford the expensive equipment required. Therefore, the difficult-to-implement JIT production system would have to be used. The plant thought briefly of changing the market qualifying and order winning outputs to emphasize flexibility and innovativeness but rejected this strategy because their market research and competitive analysis indicated that this was not what their customers wanted. Besides, the plant had never been able to make money in the past when it had provided high levels of flexibility and innovativeness. In summary, the results after step 2 were:

Market qualifying outputs—Cost, quality, performance
Order winning output—Delivery
Required production system—JIT

Step 3: What must the ABC plant do to get from where it is to where it wants to be?

ABC had to change its batch flow production system to a JIT production system. This meant that each manufacturing lever had to be moved from its current position, which was appropriate for the batch flow production system, to a new position that would be appropriate for the JIT production system. Some of these adjustments are shown in Figure 6-4.

Human Resources

Adjustments to this lever included implementing an employment security policy, starting training programs to give employees the skills they needed to participate in the change process and the skills they needed to do the new jobs in the JIT production system, starting a program to equip employees with multiple skills, negotiating a pay-for-knowledge scheme with the union so employees would be paid for the number of jobs they were qualified to do, organizing problem solving teams, improving communications by holding regular meetings, publishing a newsletter, starting a special suggestion program, and so on. Specialized skills were acquired by hiring a new process engineer and some new managers. These adjustments are represented in Figure 6-4 by the circle, with the number 1 inside, and arrows in the manufacturing levers section of the worksheet. The row on which the symbol is located indicates that this is an adjustment to the human resources lever. The columns in which the circle and arrows are located signify that this adjustment is designed to help the production system provide higher levels of the delivery, cost, and quality outputs.

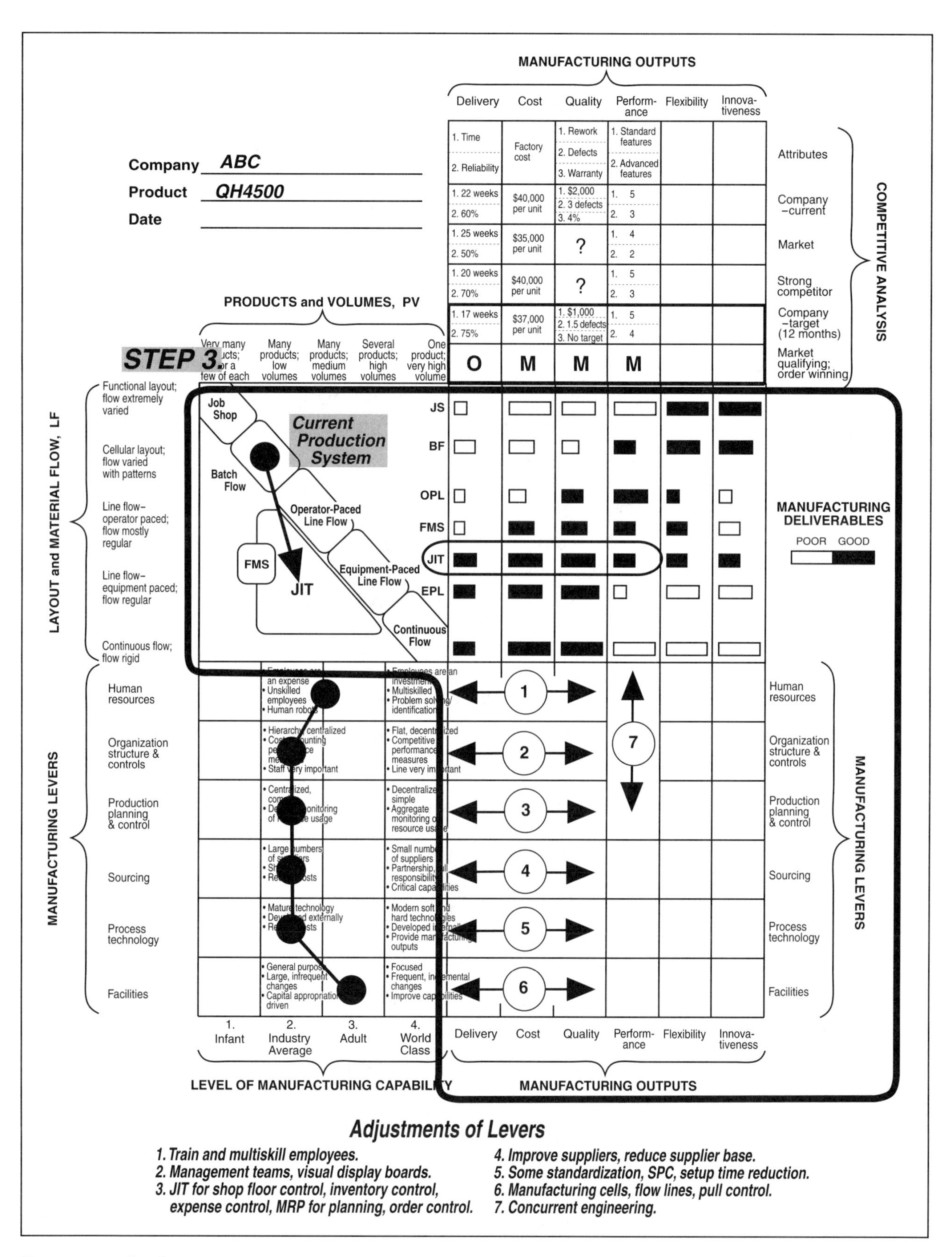

FIGURE 6–4
Step 3. Lever Adjustments Required to Change the Production System

Organization Structure and Controls

Some departments were realigned. Many supervisors were reassigned. Responsibility and decision making were pushed to the lowest possible levels of the organization. Teamwork was stressed, and bonuses were awarded on the basis of team performance. Large display boards were erected around the plant on which important information was displayed. Steps were taken to develop closer ties with the marketing group and the product design group.

Production Planning and Control

A JIT control system and many JIT techniques were implemented. For example, inventory was moved from the stockroom to the plant floor, a two-bin pull system was implemented to control production activities, lot sizes were reduced, and so on. Some standardization and modularization of components was completed. The existing computerized production planning system and cost accounting system were modified to provide better support for the JIT production system.

Sourcing

The number of suppliers was reduced. Partnerships were formed with the remaining suppliers in which quality and fast, reliable deliveries were emphasized.

Process Technology

Equipment was moved, the layout was changed, and some manufacturing cells were organized. Setup times were reduced and some equipment was improved. Some special tooling was built.

Facilities

The plant was reorganized to accommodate a JIT line flow for product family QH4500. Most of the stockroom was eliminated. Some simple, inexpensive equipment was purchased.

Once the necessary changes were determined and summarized on the strategy worksheet, they were reviewed to determine whether they could be implemented. This consisted of making two checks:

1. The following question was asked for each manufacturing lever (see Figure 6-5): Given the current level of manufacturing capability for this lever, is the organization able to effectively implement the changes that are suggested for the lever?

To illustrate, consider the human resources lever. ABC identified four major changes for this lever—training and multiskilling, a pay-for-knowledge wage scheme, team approaches, and improved communications. The level of manufacturing capability was slightly above industry average. Most organizations with this level of manufacturing capability would find it difficult to implement these four changes all at once. ABC had no choice, however, it had to make all the changes quickly. The same was true of the changes to the production planning and control lever. It would be difficult to implement the many changes that were needed—move inventory from the stockroom to the plant floor, implement a two-bin pull system, reduce lot sizes, modify the existing computer systems, and so on—when the organization has only an industry average level of capability in this subsystem. The solution was to organize all the changes into a careful implementation plan (see Chapter 7).

2. After an individual check was made of each lever, an overall assessment was made.

The ABC plant had an overall level of manufacturing capability that was slightly better than industry average. That was the foundation on which ABC would make all the changes. This capability was also the major source of resources for the implementation. Clearly, it would be difficult for the ABC plant to implement all the changes that were identified in Figure 6-4. A careful implementation plan was needed. The implementation plan specified the sequence in which changes would be made, the speed at which each change would be made, and the resources that would be required.

Summary

The five elements of manufacturing strategy presented in Chapters 1 to 5 constitute a framework that can be used to analyze manufacturing and develop plans for improving it. A three-step

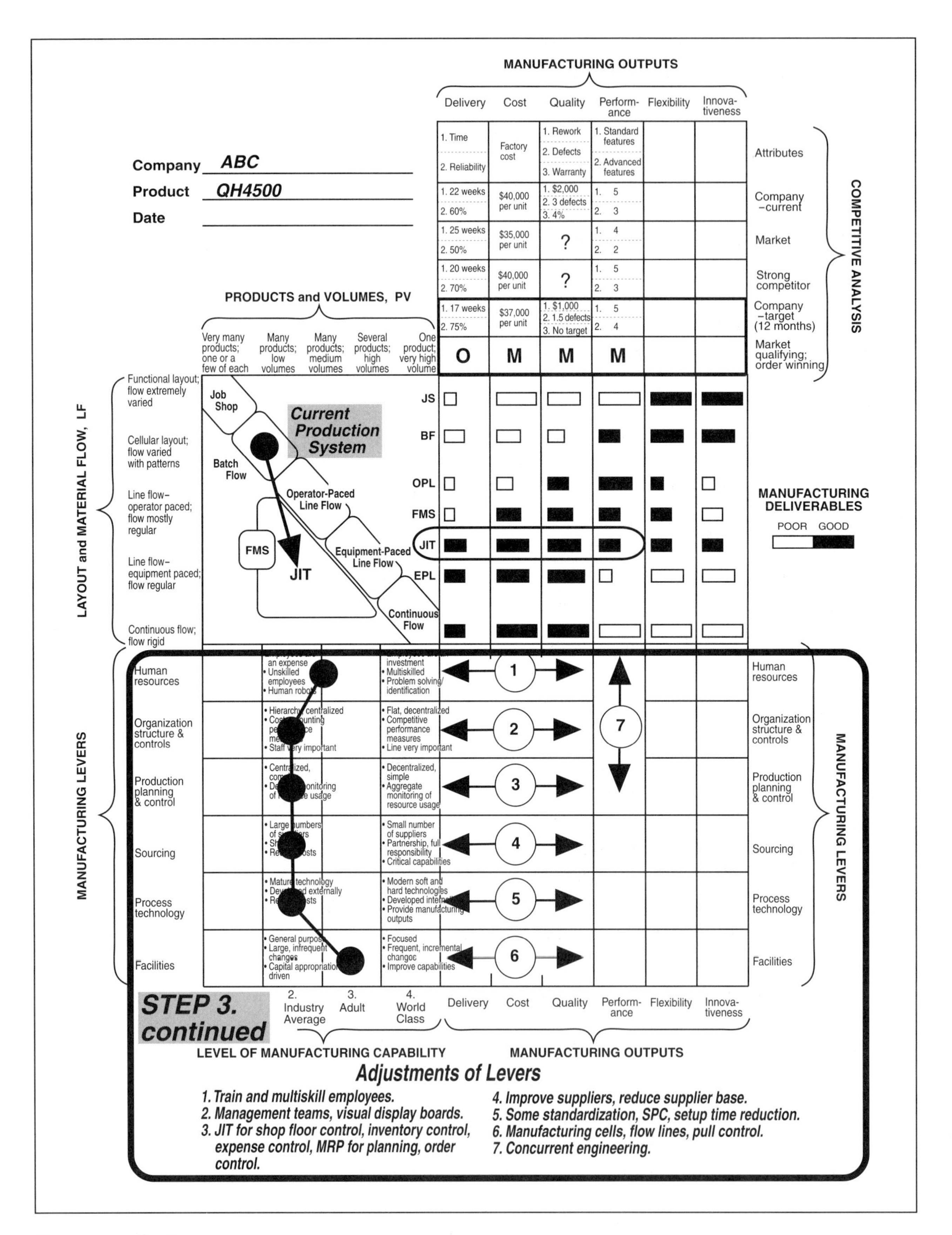

FIGURE 6–5

Step 3 (continued). Are the Adjustments to the Levers Doable?

procedure for using the framework was outlined in this chapter. The steps can be described as follows: 1) Where am I? 2) Where do I want to be? and 3) How will I get from where I am to where I want to be? An example was presented to illustrate the procedure.

Further Reading

Beckman, S. L., W. A. Boller, S. A. Hamilton, and J. W. Monroe, "Using Manufacturing as a Competitive Weapon: The Development of a Manufacturing Strategy," pp. 53–75, *Strategic Manufacturing*, P. E. Moody (ed.), Homewood, IL: Dow-Jones Irwin, 1990.

Fine, C., and A. Hax, "Manufacturing Strategy: A Methodology and an Illustration," *Interfaces*, Vol. 15, No. 6, pp. 15–27, 1985.

Appendix: Other Frameworks for Developing Manufacturing Strategy

Other frameworks for developing manufacturing strategy are in use. In Chapter 8, for example, the framework used at IBM will be described. In this appendix, we describe a framework used at Hewlett-Packard (HP). Both the IBM and HP frameworks are simpler versions of the framework presented in this book. This discussion of strategic planning at HP is adapted from an article by Beckman et al. (see the Further Reading section above). Developing a manufacturing strategy at HP consists of completing the following five steps.

Step 1

Manufacturing participates in the development of a business strategy that specifies the goals of the business, the products to be manufactured, the markets to be served, and the basis of competition for the business. The following product/market characteristics are used as measures and targets in the business strategy:

- Product variety
- Market volume
- Product standardization
- Market growth
- Rate of product change

Values of high, medium, or low are assigned to each characteristic for each product family (see Figure 6-6).

STEP 2

The product/market characteristics are translated into critical success factors for manufacturing. There are four critical success factors:

- Cost—the ability to produce a product for the lowest possible cost.
- Quality—conformance to or betterment of customer requirements for a product.
- Availability—the ability to deliver the product when and where desired, as well as the ability to respond to changes in market demand and opportunities.
- Features—the manufacturing system allows unique attributes in the product design to be included.

Values of high, medium, or low are assigned to each critical success factor for each product family (see Figure 6-6).

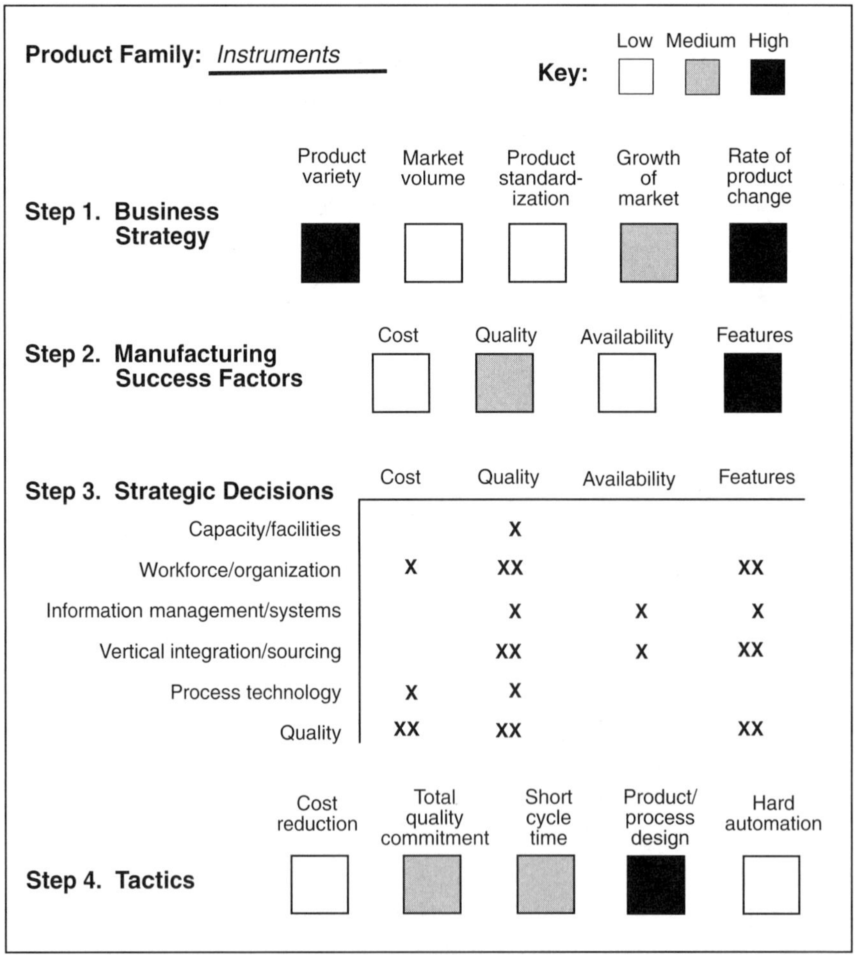

FIGURE 6–6

The Hewlett-Packard Five-Step Strategic Manufacturing Process

Product Family: *Instruments*

Key: Low Medium High

Step 1. Business Strategy

Product variety, Market volume, Product standardization, Growth of market, Rate of product change

Step 2. Manufacturing Success Factors

Cost, Quality, Availability, Features

Step 3. Strategic Decisions

	Cost	Quality	Availability	Features
Capacity/facilities		X		
Workforce/organization	X	XX		XX
Information management/systems		X	X	X
Vertical integration/sourcing		XX	X	XX
Process technology	X	X		
Quality	XX	XX		XX

Step 4. Tactics

Cost reduction, Total quality commitment, Short cycle time, Product/process design, Hard automation

STEP 3

Strategic decisions are made in six manufacturing subsystems so that the critical success factors will be provided. The six manufacturing subsystems are:

- Capacity/facilities
- Workforce/organization
- Information management/systems
- Vertical integration/sourcing
- Process technology
- Quality

The strategic decisions are summarized on a strategic decision matrix (see Figure 6-6).

STEP 4

HP identifies five tactics that can be used in the six manufacturing subsystems to provide the critical success factors:

- Cost reduction
- Total quality commitment
- Short cycle time
- Linking product design with process design
- Hard automation

Values of high, medium, or low are assigned to these tactics for each product family (see Figure 6-6).

STEP 5

Results are measured and more changes are initiated.

DISCUSSION

The similarities between the HP framework and the framework outlined in this book can be summarized as follows. Steps 1 and 2 are equivalent to the integration of manufacturing strategy with business strategy (Chapter 8), and the competitive analysis element of the manufacturing strategy framework. Steps 3 and 4 are included in two elements of the manufacturing strategy framework, production systems and manufacturing levers. HP's six manufacturing subsystems are similar to the six levers. The tactics are particular sets of adjustments to the levers. Tactics are discussed in Part II of this book: strategic thrusts and special programs (Chapters 7 and 8), and improvement approaches and soft and hard technologies (Chapters 9 and 10).

CHAPTER 7

Developing the Implementation Plan

The implementation plan is the means by which the manufacturing strategy is implemented. Information about what must be done, why it must be done, how it will be done, when it will be done, and who will do it compose the plan.

What and Why

The *whats* are the adjustments to the manufacturing levers. These adjustments or changes are summarized on the manufacturing levers element of the strategy worksheet in the columns corresponding to the market qualifying and order winning outputs most affected by the change. The *whys* are the market qualifying and order winning outputs that will be provided to the customers.

How, When, and Who

Adjustments or changes are converted into projects, and the projects are organized into a detailed implementation plan using the special worksheet shown in Figure 7-1. The plan specifies the following:

- *How:* Each adjustment to a lever is converted into one or more projects.
- *When:* The projects are prioritized and organized into a sequence for implementation. The pace at which they will be done is also specified.

- *Who:* The resources that are required to do the projects are indicated.

The Implementation Plan Worksheet

Three pieces of information from the manufacturing strategy worksheet are displayed across the top of the implementation plan worksheet: the product family, the production system, and the current and target levels of the market qualifying and order winning outputs. The main part of the worksheet is divided into rows, one for each of the manufacturing levers. The adjustments or changes to each lever are taken from the strategy worksheet and converted into one or more projects. The projects are organized into a sequence, which is displayed across the rows of the worksheet. The row for each manufacturing lever can be subdivided into smaller sections if required. For example, the row for the human resources lever can be divided into sections for

FIGURE 7–1

The Implementation Plan Worksheet

Company __________
Product __________
Date __________

	Current	Target
Production System		
Outputs – Market Qualifying		
– Order Winning		

Manufacturing Levers	Elements	Projects: Set Course	Shoot and Aim	
Human resources	• • •			
Organization structure and controls	• • •	Top management awareness, acceptance of concepts, and commitment to execute		
Production planning and control	• • •			
Sourcing	• • •			
Process technology	• • •			
Facilities	• • •			
Targets and Deliverables		• Vision, leadership, visibility, support, active participation		
Time (Months)		0 3	9	12 18 24 30

operators, support personnel, and staff. Targets and deliverables for each project are shown at the bottom of the worksheet, with the times when each project will start and finish. In addition to being organized according to the lever they affect, projects are also organized into three groups ("set course" projects, "shoot and aim" projects, and main projects); according to the times when the projects will be done.

Set Course Projects

Some simple projects are scheduled for the start of the implementation so that the implementation will start smoothly. A smooth start can create a positive environment for the rest of the implementation, build some momentum, and help personnel gain some experience. The simple projects set the course for the rest of the implementation.

Shoot and Aim Projects

At this early stage of the implementation, it is more important to get some larger, essential, visible projects started than it is to spend a lot of time planning. These projects are called shoot and aim projects because excessive aiming is not necessary. It should be obvious to everyone that these projects need to be done. The purpose of the projects is to create early, visible successes that, in turn, create momentum for the rest of the implementation, prepare a foundation for the more difficult projects that follow, and prevent "paralysis by analysis" from stalling the implementation.

Main Projects

The bulk of the projects constitutes the main part of the implementation plan. They are undertaken after the shoot and aim projects. The pace of the implementation—that is, the rate at which projects are started and the time allowed for completion of each project—depends on three factors: the urgency for improvement; the level of manufacturing capability, which is a measure of the ability of the production system to handle change; and the quantity of the resources available for the projects.

To illustrate these ideas, two actual cases are presented. In both, the need for improvement was urgent and so the implementation was done quickly. About six months was needed to develop the manufacturing strategy and the implementation plan for both cases. Two months were then taken to complete the set course projects, with another 18 months required to complete the remaining projects.

SITUATION 7.1

An Implementation Plan at ABC Company

FIGURE 7-2 shows the implementation plan for the ABC plant. It is taken from the manufacturing strategy worksheet in Figure 6-4 on page 113. QH4500 is the product family for which the strategy was developed, and it would be produced on a JIT production system. Cost, quality, and performance were the market qualifying outputs, and delivery was the order winning output. The current level and the target level for each output are also displayed at the top of the implementation plan worksheet.

SET COURSE PROJECTS

The first project in the implementation plan sought to create top management awareness and commitment. The plant felt that, without the support and active participation of top corporate management, any major effort to improve manufacturing would not succeed. Top corporate management would make sure that the cooperation of other functional areas (such as product design, marketing, human resources, and information systems) was forthcoming when needed. They would also make resources available when needed.

SHOOT AND AIM PROJECTS

Four projects were started immediately to build momentum and gain confidence. Two projects focused on infrastructural levers, human resources, and organization structure, and two focused on structural levers, process technology, and facilities. Communications were improved. The final details of the manufacturing strategy and implementation plan were completed. Simple quality improvement projects were started, and some improvements to

Company ABC
Product QH 4500
Date

Production System Just-In-Time
Outputs – Market Qualifying
– Order Winning

	Current	Target
Cost	$40,000	$37,000
Quality	$2,000, 3	$1,000, 1.5
Performance	5, 3	5, 4
Delivery	22, 60	17, 75

Manufacturing Levers | Elements | Projects

Set Course | Shoot and Aim

Human resources
- Top management awareness, acceptance of concepts, and commitment to execute
- Awareness to the rest of the organization
- Increase training and communication → Multiskill employees → (Ongoing)

Organization structure and controls
- Develop a detailed implementation plan
- Change control systems to support strategic plan
- Concurrent Engineering
- Some standardization

Production planning and control
- Pull production control system

Sourcing
- Reduce supplier base

Process technology
- Begin to improve quality
- Improve quality → (Ongoing)
- Begin to improve equipment reliability and process capability
- Improve equipment reliability and process capability
- Reduce setup times

Facilities
- Improve workplace organization
- Improve layouts

Note: At the start of each project select and measure appropriate indicators (quality, lot size, cycle time, inventory, material movement, setup time, on-time deliveries, ...). Display and track indicators on display boards.

Targets and Deliverables
- Set Course: • Vision, leadership, visibility, support, active participation
- Shoot and Aim: • Provide focus and sense of direction • Develop expertise and experience • Achieve early successes • Reduce stress, find champions

Time (Months): 0, 3, 9, 12, 18, 24, 30

FIGURE 7–2

The Implementation Plan for ABC Plant

equipment were made. Nine months later, the set course and shoot and aim projects were completed and the following achievements were realized. Top management support and participation were obtained. A detailed manufacturing strategy and implementation plan were prepared. These gave the organization a clear sense of the direction in which it was headed. The early successes in the quality improvement and equipment improvement projects created confidence and enthusiasm. Champions were found for future projects.

More targeted projects followed. Projects confined to manufacturing alone were completed by the twentieth month. By that time, manufacturing had achieved significant improvements in cost, quality, and delivery, and it felt that it had earned the right to ask other functional areas to cooperate on future projects. For example, the concurrent engineering project and the standardization project required that manufacturing, product design, and marketing cooperate to redesign the QH4500 product family for the purpose of making it easier to manufacture and reducing the number of options offered to customers. The project to reduce the supplier base required the cooperation of the corporate purchasing department.

One measure of the plant's performance during this period is shown in Figure 7-3. During the first seven months of the implementation, sales (measured as shipments from the plant) remained well below the plant's $16 million per month breakeven point while the plant struggled with the new JIT production system. By April of the first year, those who opposed the manufacturing strategy were pressing top corporate management to stop the implementation. Top management believed in the strategy, however, and their commitment kept the implementation going. A few months later, the situation began to improve. As the level of manufacturing capability rose, shipments increased and other measures of plant performance improved. The plant has been profitable ever since.

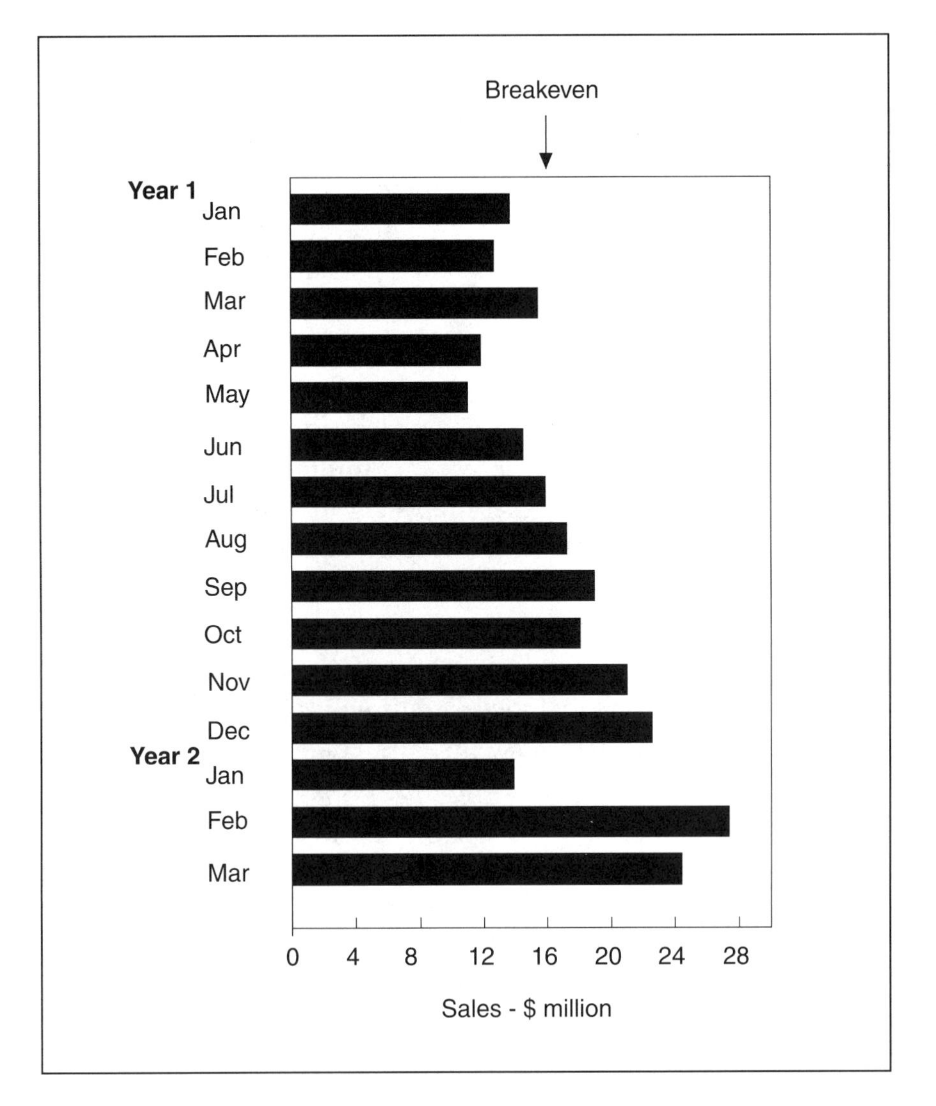

FIGURE 7–3

Shipments from the ABC Plant During the Implementation

SITUATION 7.2

An Implementation Plan at Cincinnati Milacron

THE IMPLEMENTATION plan shown in Figure 7-4 was developed from information obtained during a visit to Cincinnati Milacron's Electronics Systems Division plant and from an article by C. Powell. (See the Further Reading section at the end of this chapter.) In the 1980s, the ES Division realized that its competitors had definite cost, quality, and delivery advantages, and that it needed to improve the levels at which it provided these outputs if it was to survive in the world marketplace. After careful analysis and planning, the decision was made to focus manufacturing in three plants-within-the-plant. A new manufacturing strategy was to be implemented through the sequence of projects shown in Figure 7-4.

Subplant 1 fabricated printed circuit boards. The boards were assembled in subplant 2, and final product assembly was done in subplant 3. Manufacturing cells and line flow production lines were used in each subplant. Inventory was moved from the stockroom to the plant floor, and a pull control system was implemented. Because many JIT techniques were used (including a pull control system, multiskilled workers, and quality control), the ES Division called the new manufacturing strategy a JIT implementation. This was a little misleading because the production system was an operator-paced line flow production system that made extensive use of JIT techniques (see Chapter 17).

Other projects included a training program, a project to change the control systems, and changes to the organizational structure. The training program provided an average of 60 hours of training over two years for each employee in statistical quality control, and manufacturing, engineering, and management skills. Extensive changes were made to the MRP production planning and control system. The incentive pay scheme was eliminated. Lot sizes were reduced. The organizational structure was changed from seven reporting levels to five, and the number of job classifications was reduced from 70 to 40. Over 80 staff people were reassigned, and 2,000 square feet of office space were eliminated by moving staff personnel from production control, industrial engineering, test engineering, and process engineering to the plant floor.

By the thirtieth month, the ES Division had an operator-paced line flow production system with manufacturing capability approaching the adult level. It was able to provide its market qualifying outputs (quality, delivery, and flexibility) and order winning output (cost) at the target levels.

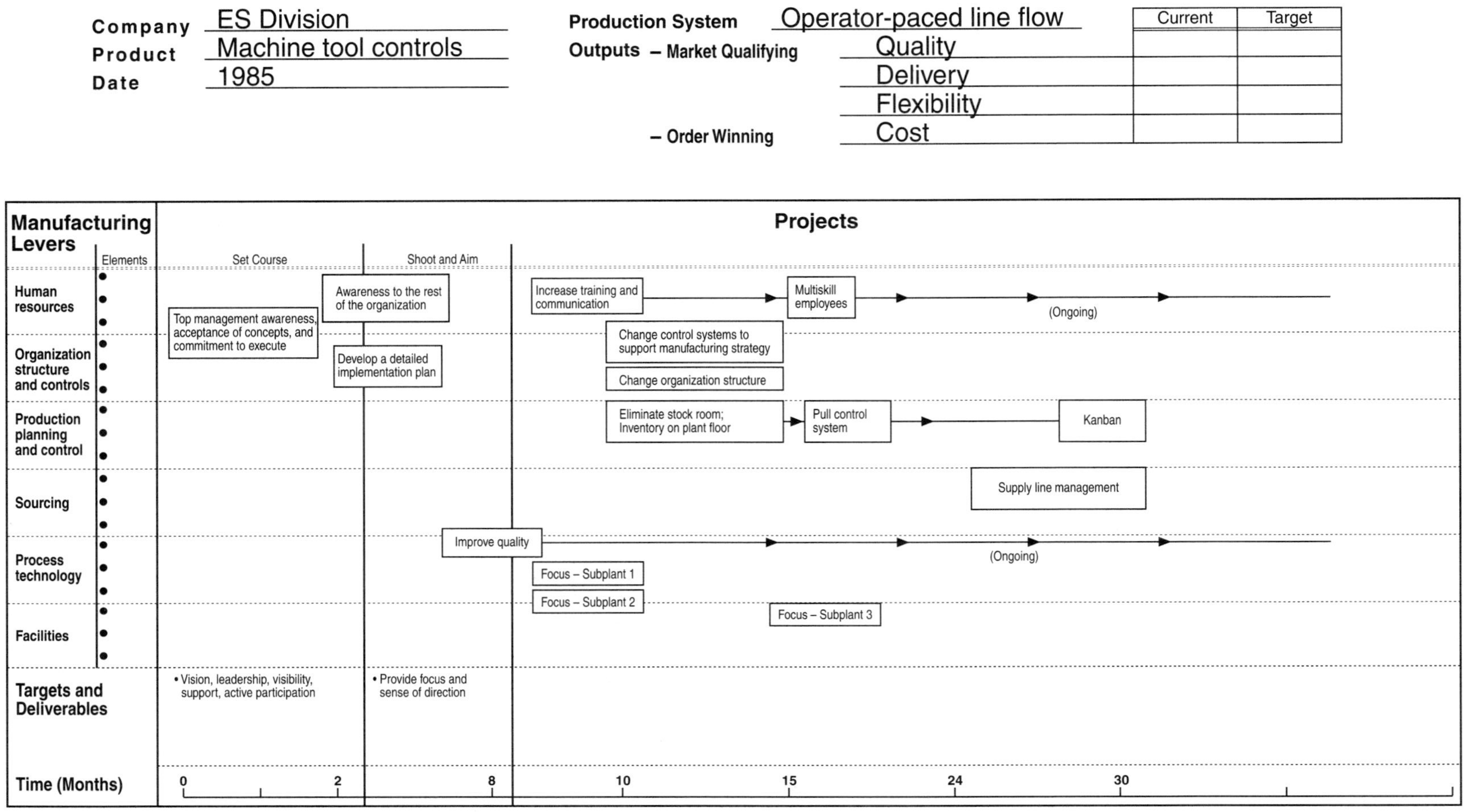

Company ES Division
Product Machine tool controls
Date 1985

Production System Operator-paced line flow
Outputs – Market Qualifying
– Order Winning

	Current	Target
Quality		
Delivery		
Flexibility		
Cost		

Manufacturing Levers
Elements
Human resources
Organization structure and controls
Production planning and control
Sourcing
Process technology
Facilities
Targets and Deliverables
Time (Months)

Projects
Set Course
Shoot and Aim
Top management awareness, acceptance of concepts, and commitment to execute
Awareness to the rest of the organization
Develop a detailed implementation plan
Increase training and communication
Multiskill employees
(Ongoing)
Change control systems to support manufacturing strategy
Change organization structure
Eliminate stock room; Inventory on plant floor
Pull control system
Kanban
Supply line management
Improve quality
(Ongoing)
Focus – Subplant 1
Focus – Subplant 2
Focus – Subplant 3
• Vision, leadership, visibility, support, active participation
• Provide focus and sense of direction
0 2 8 10 15 24 30

FIGURE 7–4

The Implementation Plan for Cincinnati Milacron, ES Division

Projects Frequently Used in Implementation Plans

The implementation plans at ABC and the ES Division contain many of the same projects. A small group of projects is found in almost all implementation plans that seek to make major improvements in manufacturing capability:

- Top management commitment and participation
- Improvements in quality, equipment, and communications
- Upgrading of skills
- Reduction in setup times
- A change in control systems to support the manufacturing strategy

Top Management Commitment and Participation

Even if not initiated by top management, major changes in manufacturing must be supported actively by this group if they are to be successful. Major changes create stresses in the organization. They require more cooperation among functional areas. Top management leadership, vision, support, and active participation are required to reduce the stress levels, ensure that cooperation takes place, and shape the organization along the lines of the manufacturing strategy.

Improvements in Quality

Quality techniques are used throughout the organization to improve quality in all activities. Statistical process control (SPC) is used in the production process to monitor quality. When necessary, quality is improved and personnel are trained to monitor the quality of their work. Benefits of this project are that, when other projects make changes, the effects on quality will be seen immediately, and should quality improve or diminish, appropriate actions can be taken. Improving the quality of materials purchased from suppliers is also included in this project. In many companies, purchased materials make up more than 50 percent of the cost of goods sold. Efforts commensurate with the importance of this part of the business are expended on improving the quality of purchased material. Buyers, materials personnel, and production planning personnel are trained, and supplier certification programs are started.

IMPROVEMENTS IN EQUIPMENT

Once quality improvement activities from the previous project are under way, improving the reliability of production equipment begins. Teams consisting of personnel from production, maintenance, tooling, materials, and process engineering study the operations required to produce each product. The process capability of each operation for each important quality attribute is determined (see Chapter 10). The layout is updated to show the locations of all materials, tooling, equipment, and operators. Standard operations sheets are prepared describing in detail how the operation is to be done, what tools and fixtures are to be used, and how and when quality checks are to be done. A maintenance schedule is prepared outlining the maintenance that should be done, the frequency with which it will be done, who will do it, and how it will be recorded. The team may find that equipment and tooling need to be improved, maintenance needs to be improved, operators need training or disciplining, and quality improvements are necessary. Appropriate actions would then be taken.

IMPROVEMENTS IN COMMUNICATIONS

Employees must be prepared for the changes that will take place by being given information in advance and by receiving training. Communications should inform all personnel of planned changes and explain why they are necessary. The goals are to gain support for the changes, and to involve employees in the change process. Four types of communications are used.

Display Boards

Large, attractively designed, clearly visible display boards, each with a specific purpose, are located throughout the facility. Some boards display information about efforts to improve quality—SPC charts, repair costs, warranty costs, and so on. Others display production information—status of orders currently manufactured, promised delivery dates, inventory levels, and manufacturing cycle times. Boards can be used to display information about equipment process capability, standard operations, equipment downtime, and repair times.

Newsletter

A high-quality newsletter may be used. The topics covered include the organization's business plans, financial status, new orders, planned changes, information on training programs, and activities in other organizations.

Meetings

Department meetings and facility-wide meetings may be held. Topics addressed include the manufacturing strategy and its implications for each department, business plans and financial conditions, status of current and upcoming changes, training programs, and so on.

Open House

Once each year, the organization should open its doors to employees and their families, as well as suppliers, customers, and others in the community. This allows those with an interest in the facility to see for themselves the changes that are being made. Some restrictions apply, however, when production involves special or proprietary practices and processes that, if leaked to suppliers and customers, could result in a loss of competitive advantage.

Upgrading Skills

Employees must be provided with two types of skills: skills to participate in the change process and skills to do the new jobs after the changes are made. Training is given in the following areas, with the amount varying according to employee need.

Quality Techniques

Training in quality techniques is given to all employees so that quality can be improved in all activities. Some employees are trained in advanced quality techniques so that additional improvements can be made.

Industrial Engineering

Many employees require training in basic industrial engineering concepts such as equipment layout, process and flow, time standards, and maintenance procedures. One benefit of this training is that it provides a common language and a common set of tools for those involved in the changes.

Setup Time Reduction, Maintenance, and Equipment Repair

Equipment used in the production process should be capable of consistently producing good products and running for long periods without breakdowns. This requires responsive maintenance and tooling departments, and production operators who are involved with setups, routine maintenance, and small repairs. Production support departments often resist these requirements because they change the nature of work in the departments. Training and careful planning are needed to overcome this resistance.

Reduction in Setup Times

Further improvements can be made to the production system through a program of setup time reduction. The benefits of short setup times include the ability to produce in small lots, reduced setup costs and inventory costs, the flexibility to produce new products quickly, and better workplace organization.

A Change in Control Systems to Support the Manufacturing Strategy

The organization's control systems are analyzed to determine whether they support and encourage achievement of the manufacturing strategy. These systems include the accounting system, the capital appropriation system, the procedure for selecting suppliers, systems for evaluating individual and department performance, the procedure for bidding on new orders, compensation systems, and so on. In many organizations, these systems were developed for a different manufacturing strategy. They may even be inconsistent because the behavior encouraged by one system is different from that encouraged by another. For example, the accounting system may discourage a department from working to improve quality, which it must do to win new business.

Special Programs

Many manufacturers organize small sets of complementary projects into programs (see Chapter 8). In 1988, Roth and Miller (1990) identified 39 different programs and, as part of a larger study, asked 193 executives from large U.S. manufacturing firms whether or not each of the 39 programs "was currently being given a significant degree of emphasis in order to improve operations effectiveness." Twenty-eight of the programs were identified as receiving a significant degree of emphasis (see Figure 7-5). Roth and Miller organized the 28 programs into two categories, structural and infrastructural, depending on which part of the organizations' production system they had the largest effect. They also organized the programs into seven types. Notice that there is a rough correspondence between these types and the six manufacturing levers.

There are two issues that the study did not explore. 1) Were there more specific objectives, beyond simply improving operations effectiveness, that each organization sought to achieve when it implemented its programs? That is, were there specific manufacturing outputs—cost, quality, performance, delivery, flexibility, and innovativeness—that organizations were trying to improve? 2) Into what production systems were the organizations implementing the programs? The programs are not necessarily appropriate for all production systems. Despite these limitations, the study shows that sets of complementary projects or programs are widely used in manufacturing strategy.

Monitoring Projects During the Implementation Plan

The implementation plan worksheet in Figure 7-1 organizes projects according to the manufacturing levers they affect. It may also be useful to organize projects according to the manufacturing outputs they affect. The form shown in Figure 7-6 is used to do this. The four projects in the figure sought to improve quality at the ABC plant. The projects were prioritized, and the financial and labor resources required to complete each project were identified. Responsibility for each project was assigned, and completion dates were set.

It is useful to track the progress of each project on a display board so that all employees can follow the project. At the start of each project, appropriate indicators, such as the values of quality attributes, repair costs, inventory levels, lot sizes, cycle time, conveyor length, setup time, percent of on-time deliveries, and open customer orders, are selected and measured. The indicators should improve as changes are made. The improvements, however small, are posted on the display board so that they do not go unnoticed. In

FIGURE 7–5

Some Important Manufacturing Programs

Structural or Infrastructural	Type of Program	Program
Structural	Material flow	Manufacturing lead time reduction Supplier lead time reduction Just-in-time Reducing setup/changeover time
	Advanced process technology	Flexible manufacturing systems Group technology Robots Computer aided design
	Capacity upgrade	Capacity expansion Reconditioning of physical facilities Automating jobs
	Restructuring	Reducing size of manufacturing units Plant relocation Manufacturing reorganization Closing plants Changing labor-management relationships
Infrastructural	Resource improvements	Supervisor training Management training Worker training Preventive maintenance Job enrichment Worker safety
	Quality improvements	Statistical quality control — product Statistical process control — process Supplier quality Zero defects
	Information and systems	Integrating systems across areas Integrating systems within manufacturing

Source: 1988 data from Roth and Miller 1990.

Company **ABC Company**
Product **QH4500**
Manufacturing Output **Quality**

Project	Priority	Cost	Labor Requirements	Scheduled Completion	Responsibility
Improve quality training	Top priority	$150,000	2 employee-years	August, year 1	Human resource manager
Determine process capabilities and develop process controls	Highly desirable	$80,000	1 employee-year	End of year 2	Engineering manager
Improve tooling and fixtures	Desirable	$125,000	1½ employee-years	End of year 1	Engineering manager
Fit quality into incentive wage scheme	Desirable	$60,000	½ employee-year	End of year 1	Plant manager

FIGURE 7–6

Organizing Projects by Manufacturing Output

this way, enthusiasm is sustained and momentum builds. Tracking appropriate indicators also helps keep projects on course.

Gaining the support and participation of employees is necessary for manufacturing change to be successful. But it cannot be gained overnight. In many plants, an adversarial manager-worker relationship, developed over the years, must be overcome. Three favorable situations can help gain this support: 1) Improvements achieved from earlier projects are used to demonstrate the importance of the new manufacturing strategy and its implementation plan, 2) information is communicated quickly to all employees so that there are no surprises or misunderstandings, and 3) the issue of job security is resolved. Employees do not feel that their livelihood will be affected adversely by their participation in the implementation plan.

SUMMARY

The last step in developing the manufacturing strategy is determining the adjustments that will be made to each manufacturing lever. These adjustments are converted into projects, which are prioritized and organized into a detailed implementation plan using the implementation plan worksheet. The worksheet includes the following information:

- The product family under consideration

- The production system to be used
- The current and target levels of the market qualifying and order winning manufacturing outputs
- Set the course projects: These simple projects are designed to start the implementation on a solid basis.
- Shoot and aim projects: These are obvious, necessary projects that seek to produce early, visible successes and prepare a foundation for the more difficult projects to follow.
- Main projects: These projects make most of the changes and generate most of the improvements. They are the adjustments to the manufacturing levers.

Projects that are found frequently in all implementation plans were discussed. The pace of an implementation depends on how critical the need for improvement is, the level of manufacturing capability, and the quantity of resources available for the projects.

Further Reading

Powell, C., "Cincinnati Milacron—Electronic Systems Division: Implementing JIT for Survival in the Machine Tool Market," *Target*, Vol. 3, No. 2, pp. 28–32, Summer 1987.

Roth, A. V., and J. G. Miller, "Manufacturing Strategy, Manufacturing Strength, Managerial Success, and Economic Outcomes," *Manufacturing Strategy*, J. E. Ettlie, M. C. Burstein, A. Feigenbaum (eds.), Dordrecht, The Netherlands: Kluwer Publishers, pp. 97–108, 1990.

Appendix: Project Guidelines During the Implementation Plan

The following guidelines are cited frequently as ways to increase chances for success and reduce levels of stress when implementing major changes in manufacturing:

- High visibility and leadership are required from top management.
- Carefully chosen champions are needed.
- Initial expectations may be too high. Implementations most often move more slowly than planned. Unforeseen problems should be expected.

- Some groups will have difficulty developing plans. They will proceed hastily and make mistakes.
- Know the production process. Take the time to carefully study the current process so that everyone is working with reality.
- Pay attention to details.
- Be certain that those who contribute to the plan are those who have to live with the changes after they are implemented.
- Groups who lose influence will feel threatened, and groups who gain influence will support change.
- Different functional areas may need encouragement to cooperate with each other.
- The engineering department is often asked to "put away its hi-tech toys and get out on the plant floor to implement low-tech solutions." This instruction will be unpopular.
- Profitability is the concern of top management, but middle managers will be concerned about loss of influence. Workers worry about training and the security of their jobs.
- There will be some managers and workers for whom old work habits are so ingrained that they cannot change. These personnel should be transferred to other parts of the organization.
- Reward and performance measurement systems must change or employees will revert to their old habits.
- It is difficult to overestimate the amount of training that will be needed.

CHAPTER 8

INTEGRATING MANUFACTURING STRATEGY WITH BUSINESS STRATEGY

Manufacturing is not the only part of the organization that formulates and implements strategic plans. The relationship between manufacturing strategy and the strategic plans of other parts of the organization is the subject of this chapter. A large organization that is involved in many businesses may allow each business to operate differently as long as they all support the overall objectives of the organization. To help manage this, a hierarchy of businesses and strategies is defined (see Figure 8-1). Functional areas such as manufacturing, marketing, and distribution make up a business unit. Business units make up a corporation, and corporations make up an industry. Strategic plans are formulated at each level in the hierarchy, starting at the top, and are fed downward. Plans made at higher levels provide the boundaries within which plans at lower levels are made. Lower level plans are fed upward, resulting in consistent competitive plans everywhere in the hierarchy.

INDUSTRY STRATEGY

Strategy at this level focuses on issues such as incentives for investment; import and export practices; government policies; and infrastructure such as transportation, communications, education, and health care.

FIGURE 8–1

A Business Strategy Hierarchy

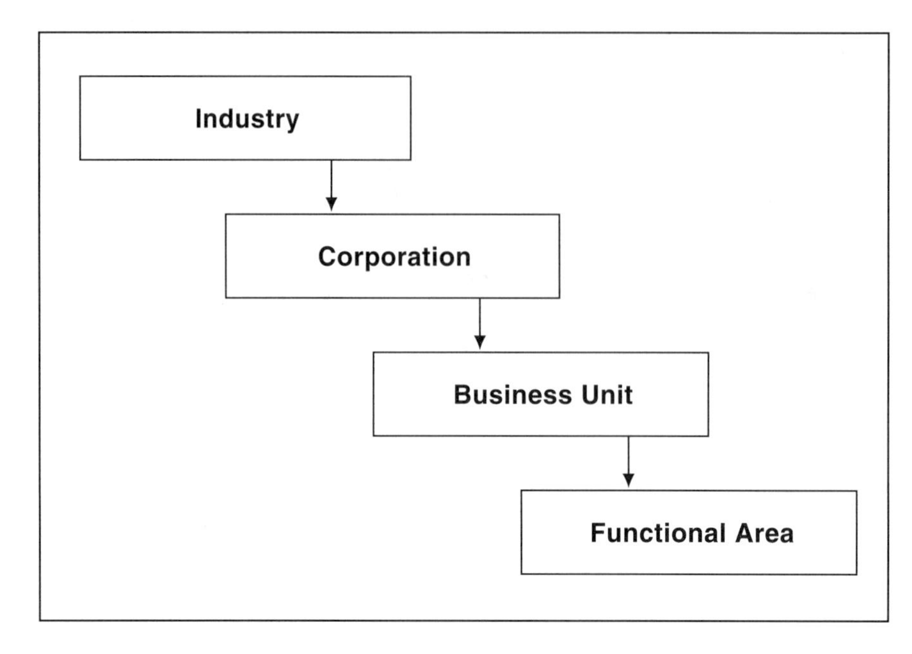

CORPORATE STRATEGY

Corporate strategy is the long-term, general statement of goals, product-market domain, and the basis of competitive advantage for the corporation. It specifies the businesses in which the corporation will participate and how resources will be allocated among these businesses.

BUSINESS UNIT STRATEGY

An organization at this level is called a strategic business unit. The business unit strategy is more short-term in outlook than the corporate strategy, and it contains a more detailed statement of the goals, product-market domain, and basis of competitive advantage for the business unit. Each business unit strategy must support the corporate strategy.

FUNCTIONAL STRATEGY

Each functional area—manufacturing, marketing, distribution, research and development, and so on—develops a strategy that outlines how it will help achieve the business unit strategy. Functional strategies must complement each other. The framework outlined in this book is used to develop the manufacturing strategy. The competitive analysis element of manufacturing strategy is also an element of the business unit strategy because the market qualifying and order winning outputs for each product family are the basis of competitive advantage for the manufacturing function and for the business unit.

ELEMENTS OF BUSINESS STRATEGY

It is useful to think of business strategy, whether it is corporate strategy or business unit strategy, as consisting of three elements: goals, product-market domain, and basis of competitive advantage.

GOALS

Businesses pursue a mix of hard and soft goals (see Figure 8-2). Hard goals are based on traditional measures of financial performance. Soft goals relate to what the business unit, as a social entity, wishes to achieve. The particular goals sought in a business strategy depend on the business's internal environment, and the opportunities and challenges in the external environment. Situation 8.1 gives an example of a statement of corporate goals.

FIGURE 8–2

Business Strategy Goals

Goal	Measures
Hard Goals	
• Profitability	– Return on sales, return on net assets, return on equity – Earnings per share
• Market Position	– Market share – Rank in industry – Diversity of product line
• Growth	– Increase in sales, assets, earnings – Increase in earnings per share
• Risk	– Liquidity – Ratio of debt to equity – Fixed charge coverage
Soft Goals	
• Management	– Autonomy – Status
• Employees	– Economic security – Opportunities to advance – Working conditions, quality of working life
• Community	– Control of externalities – Contribution to welfare, cultural life
• Society	– General benefits through innovation and efficiency – Preservation of environment – Responsible political involvement

Source: Adapted from Fry and Killing 1988.

SITUATION 8.1

Hewlett-Packard's Statement of Corporate Goals

Hewlett-Packard: Statement of Corporate Goals[1]

HARD GOALS

- *Profit:* to achieve sufficient profit to finance our company growth and to provide the resources we need to achieve our other company objectives.
- *Growth:* to let our growth be limited only by our profits and our ability to develop and produce technical products that satisfy real customers' needs.
- *Customers:* to provide products and services of the greatest possible value to our customers, thereby gaining and holding their respect and loyalty.
- *Fields of interest:* to enter new fields only when the ideas we have, together with our technical, manufacturing, and marketing skills, ensure that we can make a needed and profitable contribution to the field.

SOFT GOALS

- *Our people:* to help Hewlett-Packard people share in the company's success, which they make possible; to provide job security based on their performance; to recognize their individual achievements; and to ensure the personal satisfaction that comes from a sense of accomplishment in their work.
- *Management:* to foster initiative and creativity by allowing the individual freedom of action in attaining well-defined objectives.
- *Citizenship:* to honor our obligations to society by being an economic, intellectual, and social asset to each nation and each community in which we operate.

Goals should be consistent with each other. For instance, consider the following hypothetical business unit whose goals are neither consistent with each other nor compatible with the production system used by the business. XYZ Company uses an equipment-paced line flow production system to manufacture products in the mature stage of their product life cycles. Cost is the order winning output. XYZ's business unit goals are to double sales, increase profit margins, and maintain an entrepreneurial climate in the business. The problems with these goals are many. It is unlikely that sales of a mature product can be doubled, or that the profit margin for a mature product competing on the basis of cost can be increased much. It is difficult for an equipment-paced line flow production system to be entrepreneurial.

Product-Market Domain

This element of business strategy defines the environment in which the business competes. The environment can be depicted on a product-market matrix, where products are organized into product lines on the basis of manufacturing or market factors, and markets are organized into market segments on the basis of customers, distribution channels, competition, etc. Information such as annual sales, market share, profitability, and assets employed, which are useful for strategic analysis, is shown in each cell of the product-market matrix. Situation 8.2 on page 144 shows a product-market matrix.

Basis of Competitive Advantage

When an organization operates in a competitive environment, it needs a sustainable competitive advantage. Without one, it operates at a disadvantage. Manufacturing's competitive advantage comes from providing the manufacturing outputs—cost, quality, performance, delivery, flexibility, and innovativeness—at market qualifying and order winning levels. Some of these advantages require the participation of other functions. For example, performance requires close cooperation between manufacturing and the product design function. A business unit can acquire a basis of competitive advantage from its manufacturing function by emphasizing the competitive advantage provided by manufacturing. In addition to these outputs, the business unit can also provide complementary nonmanufacturing outputs, such as prepurchase assistance, postpurchase service, guarantees, financing and distribution, at market qualifying and order winning levels. The business strategy is a clear statement of what manufacturing and nonmanufacturing outputs will and will not be provided.

SITUATION 8.2

The Product-Market Matrix for a Pipe Manufacturer

A BUSINESS unit of a large company produced a variety of plastic pipes for use in drainage applications. The pipes were produced in five different plants. The product-market matrix shown in Figure 8-3 summarizes the way the business unit viewed its environment. Products were organized into product lines according to the production system used to manufacture them. The market was divided into three segments on the basis of the type of customer.

FIGURE 8–3

A Product-Market Matrix

Product Line	Produced in Plants	Market Segment: Agricultural • Major customers are contractors	Industrial • Major customers are supply dealers	Municipal • Direct selling	Overall	Major Competitors
4 inch pipe • Continuous flow production system • Mature product, commodity • High volumes	A, B C, E	$21 M 12%	$9 M 15%	none none	$30 M 12.7%	Small producers of plastic and clay pipes
6 to 14 inch pipe • Batch flow production system • Medium volumes • Special dies required	A, B D	$6 M 24%	$1 M 30%	$2 M 25%	$9 M 24.8%	
14 inch and larger pipe • Operator-paced line flow production system • Produced on special equipment • Low to medium volume	D, E	$2 M 60%	$1 M 40%	$6 M 65%	$9 M 59.8%	Producers of galvanized steel pipes
Overall		$29 M 14.3%	$11 M 16.7%	$8 M 46.4%	$48 M 16.1%	

Note: Each cell contains annual sales (millions of dollars) and market share (percent).
Source: Adapted from Fry and Killing 1988.

One mistake sometimes made during the development of business strategy occurs when business managers assume that there are only two ways to achieve a sustainable competitive advantage—low cost or differentiation. (See Situation 8.3.) *Low cost* requires that products be manufactured at the lowest possible cost in the industry. *Differentiation* means providing a product that is different from competitors' products. When the term *differentiation* is used in business strategy, it mixes four different manufacturing outputs—performance, flexibility, innovativeness, and delivery. Combining manufacturing outputs in such a coarse way can cause an organization to overlook important strategic options. Each manufacturing and nonmanufacturing output should be considered individually.

SITUATION 8.3

A Traditional View of Competitive Advantage

HRT COMPANY'S corporate strategy describes the basis of competitive advantage for all their businesses as follows:

> The businesses within the Company are expected to achieve and maintain a leadership position in attractive industries. A true leadership position means having a significant and well-defined advantage over all competitors. This can be achieved through continuous, single-minded determination to achieve one or both of the following positions within an industry:
>
> **Lowest Delivered Cost Position**
>
> A business with the lowest delivered cost has greater economies of scale than other competitors. Economies of scale are available in the manufacturing, distribution, and installation steps where large amounts of costs are incurred...
>
> **Differentiated Products**
>
> Differentiated products are those that offer the customer some important and unique benefit. Typically, patents, trademarks, brand names, or specialized skills prevent competitors from copying such products. If a product is truly differentiated, the customer is selectively insensitive to price. Increasing customer price sensitivity is a sign that a product is losing its differentiation advantage.[2]

Corporate Strategy

Corporate strategy is the first strategy to be set. It seeks to answer the question, "What businesses should we be in?" Goals are long-term and aggregate, and are described using dimensions such as return on sales, return on equity, rank in industry, and growth. The product-market domain defines the portfolio of businesses that comprise the corporation. A general description of the basis of competitive advantage is given. A corporate strategy may describe the relative importance of particular businesses and outline in general terms the corporate resources that will be committed to them.

Business Unit Strategy

Loosely defined, a business unit or strategic business unit is a part of an organization that satisfies most of the following four criteria: 1) It has external customers; 2) it has external competitors; 3) it is relatively autonomous because it can decide what products to produce and how to produce them, and it can select suppliers and decide how to market its products; and 4) it is a profit center. The big question in business unit strategy is, "How are we going to compete?" That is, what are the market qualifying and order winning manufacturing and nonmanufacturing outputs for each product?

Small organizations that are uncomfortable with strategic planning can use the framework presented in this book for all their strategic planning. First, they develop a manufacturing strategy for each product family. These strategies are combined into an overall strategy that becomes the strategic plan for the entire business unit. The strategies for the other functional areas follow from this business unit strategy.

Integrating Strategies

A simple, six-step process for integrating strategies from the corporate, business unit, and functional areas is shown in Figure 8-4.

Step 1: *Formulate corporate strategy.*

Step 2: *Formulate business unit strategies.*

Step 3: *Formulate functional strategies.* Business unit

strategies are fed down to the functional areas—manufacturing, marketing, distribution, research and development, and so on. Each functional area then develops a functional strategy. These strategies specify how each functional area will support the business unit strategy.

Step 4: *Consolidate functional strategies.* The functional strategies are reviewed and revised to ensure that they are consistent with each other and that they support the business unit strategy.

Step 5: *Consolidate business unit strategies.* The business unit strategies are reviewed and revised to ensure that they are consistent with each other and that they support the corporate strategy.

Step 6: *Begin implementation of the strategic plans.* After all the strategies have been developed, implementation begins. Implementation plans specify how and when the corporate strategy, the business unit strategies, and the functional strategies will be achieved.

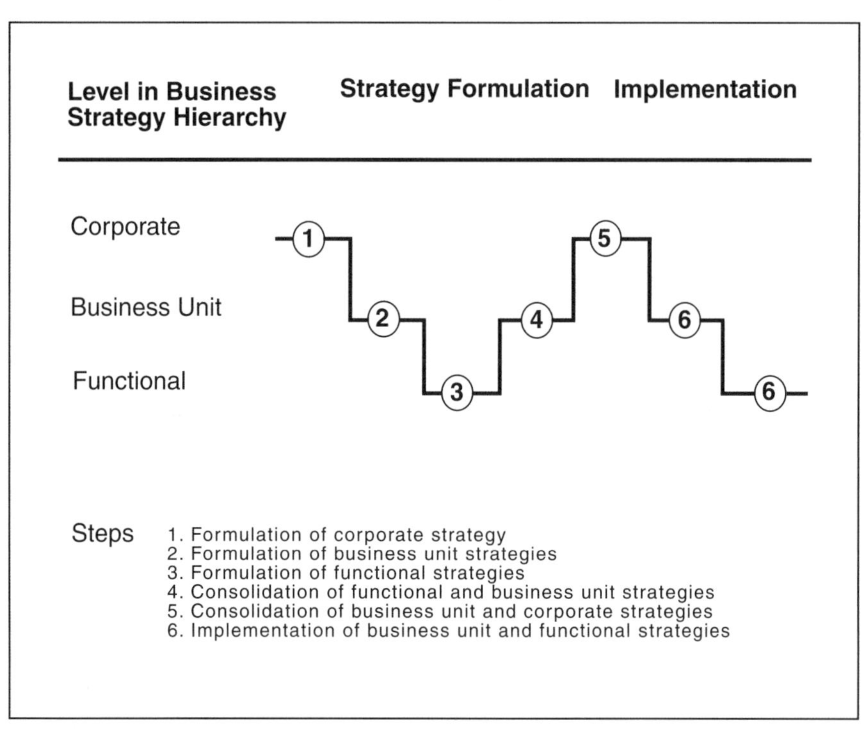

Source: Hax and Majluf 1984.

FIGURE 8–4

A Process for Integrating Business Strategy

Let's look at how these steps were followed at IBM.

SITUATION 8.4

Integrating Manufacturing and Business Strategy at IBM[3]

IN THE EARLY 1980s, IBM changed its corporate strategy and committed itself to becoming the best in the world in manufacturing. It followed an approach called "high-volume, low-cost manufacturing." Two important elements in this approach were 1) raise the level of manufacturing capability to a world class level, and 2) change the production systems to line flow production systems to provide the highest possible levels of the cost and quality outputs (see Figure 8-5). IBM also changed the process it used to develop strategy to one that closely followed the six steps listed in this chapter.

In 1985, before the company changed its corporate strategy, manufacturing at its RTP facility could be described as follows:

- Over 250 products consisting of more than 60,000 parts were developed, programmed, and manufactured.
- More than 250 machine types were used.
- Products had many options.
- Production was make-to-order.
- Schedule changes were frequent.
- The variability in the number of products produced and the production volumes was high.
- The manufacturing process was complex.
- Many products were very difficult to produce.
- The production planning and control systems were complex.

Then the corporate strategy changed and the new process for developing strategy was started.

STEP 1: CORPORATE STRATEGY

High-volume, low-cost manufacturing was the approach IBM used to become the best in the world in manufacturing.

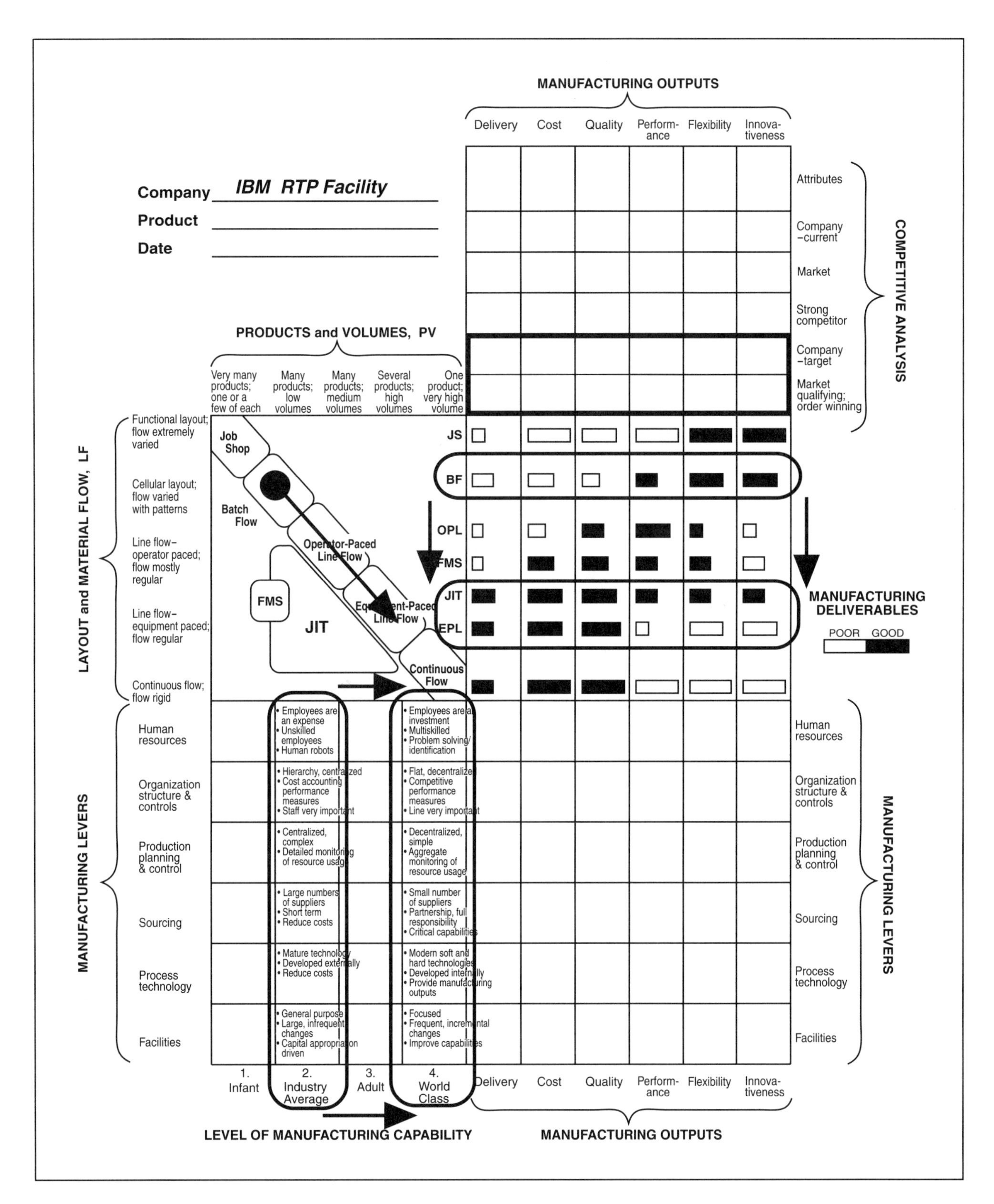

FIGURE 8–5

Manufacturing Strategy at IBM's RTP Facility

Step 2: Business Unit Strategy

The RTP facility was organized into business units, each of which developed a business unit strategy. The strategy included the unit's goals relative to measures such as cost (and profitability), quality, performance, delivery, flexibility, and innovativeness (which included growth for the products produced by the business unit). Market qualifying and order winning outputs were selected, and benchmarking was used to set targets.

Step 3: Functional Strategies

Manufacturing, engineering, marketing, distribution, financing, service, and so on developed functional strategies consistent with the business unit strategy.

Step 4: Consolidation of Functional Strategies

IBM called this step "strategic sizing." The functional strategies were summarized and compared to each other. The results of the comparisons were fed back to the functional areas so that appropriate revisions could be made. This iterative process stopped when all the functional strategies were consistent with each other and with the business unit plan. The results at RTP were the following functional strategies.

Manufacturing

Only those production systems located in the lower part of the PV–LF matrix—equipment paced-line flow, continuous flow, JIT and FMS—would be used at RTP because they were the only systems capable of providing the market qualifying and order winning outputs at the target levels. The number of products would be reduced and the volumes increased so that these production systems could be used. And there were other changes. Plants-within-the-plant (PWPs) were organized so that products could be manufactured on dedicated FMS, JIT, equipment-paced line flow, or continuous flow production systems. Robotics-based automation was installed. Products were standardized. Production was make-to-plan. Schedules were stabilized. Manufacturing and suppliers were given up to three months of frozen schedule. Finished goods inventory was maintained at the distribution center so that orders could be delivered within one week.

Engineering

Engineering strategy focused on standardizing products and designing products that were easy to manufacture.

Marketing

Marketing strategy focused on marketing standard products (rather than products with options) with competitive prices, high quality, and a one-week delivery. Marketing also sought to increase sales volumes because higher volumes were needed to fully utilize the new production systems.

Distribution

The distribution strategy concentrated on maintaining finished goods inventory and delivering products to customers within one week of receipt of a customer order.

Finance

The finance strategy included making funds available for implementing the changes in manufacturing, engineering, marketing, and distribution. New costing methods were developed for the PWPs so that more accurate cost data would be available. Benchmarking was used to develop cost targets against which actual costs would be tracked.

While the functional strategic plans were developed, two-year implementation plans were also written.

Strategic Thrusts and Special Programs

A corporation may identify a strategic thrust or direction it wants the entire organization to follow. For example, there may be an important issue that the organization needs to address over the next three to five years to improve its competitive position. Examples include developing new products, reducing cycle times, and increasing participation in improvement activities. Strategic thrusts will have different consequences for business units and the functional areas in each. Special programs are responses by manufacturing to strategic thrusts. A special program usually affects several production systems, and it can be thought of as a group of adjustments to manufacturing levers in many production systems. Some popular special programs were listed in Figure 7-5 on page 135.

Summarizing Strategies

Large corporations comprise many business units, each of which may have a different strategy. Each produces many different product families, and each product family has its own manufacturing strategy. A tool is needed to organize all the manufacturing strategies into a framework that is convenient for study and analysis. The attractiveness-strength-contribution graph is such a tool (see Figure 8-6). It is used to summarize the strategic roles of the production systems in a business unit. It has three dimensions—the attractiveness of the market, the strength of the production system, and the contribution of the production system to the business unit. Market growth rate for the products produced by the production system is used as a measure of the attractiveness of the market and is placed on the vertical axis of the graph. The relative market share is used as a measure of the strength of the production system. Relative market share is the market share held by the products produced by the production system expressed as a percentage of the market share of the strongest competitor in the industry. This measure is placed on the horizontal axis. A measure such as sales, profits, or return on assets is used for the contribution of the production system to the business unit. This measure is shown as a circle of varying diameter on the graph. Each production system is represented by a circle on the graph. The position of the circle depends on the

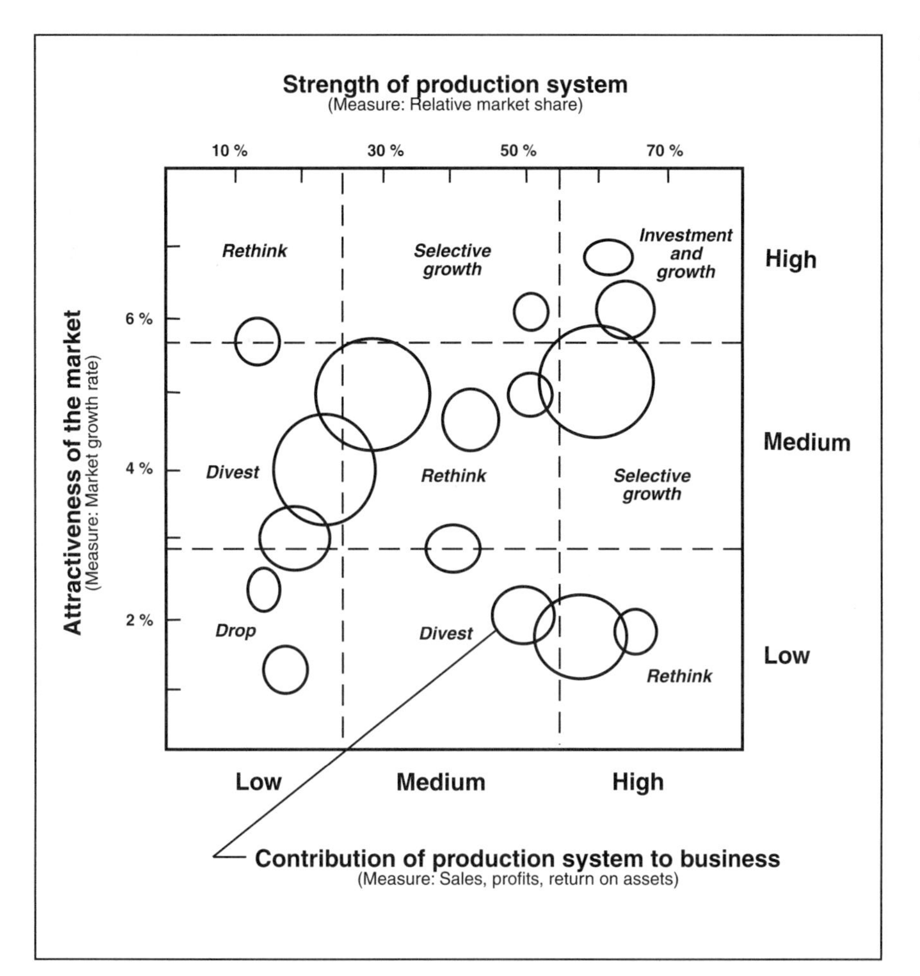

FIGURE 8–6

The Attractiveness-Strength-Contribution Graph

values of the first two dimensions. The size of the circle depends on the value of the third dimension.

The graph is divided into regions, each of which has an implied strategic direction for the production systems located there. For instance, when measures of market attractiveness, strength of the production system, and contribution of the production system to the business unit are high (large circles in the upper right corner of the graph), the natural strategy is to invest in the production system. At the other extreme, when measures of all three dimensions have low values that cannot be improved (small circles in the lower left corner), the natural strategy is to drop the production system. When measures of attractiveness, strength, and contribution have intermediate values, the business unit should think carefully about how it assigns resources to the production system.

More complex measures can be used for each dimension. For example, the attractiveness of the market can be measured by analyzing factors including market size, market growth rate, competitive structure, barriers to entry, industry profitability, technology, regulation, labor availability, and environmental issues. The strength of the production system can be measured by analyzing factors such as market share, level of manufacturing capability, levels of the market qualifying and order winning outputs, sales force, customer service, research and development, distribution, financial resources, and image.

The attractiveness-strength-contribution graph is traditionally used to summarize the strategic roles of business units in a corporation (see Hax and Majluf 1983). In this case, the three dimensions are attractiveness of the markets in which a business competes, internal strength of the business unit, and contribution of the business unit to the corporation.

SUMMARY

Strategies are formulated and implemented at many levels in an organization. A business strategy hierarchy and a six-step process for formulating strategy at each level in the hierarchy were outlined. All strategies in the hierarchy are interdependent. The strategic plans made at higher levels are fed downward and provide the boundaries within which strategic plans at lower levels are made. Business strategy consists of three elements: goals, product-market domain, and basis of competitive advantage. These three elements are major inputs to the competitive analysis element of manufacturing strategy.

NOTES

1. Excerpts from the Hewlett-Packard *Corporate Objectives* document (No. 5957-2130 7/89 Hewlett Packard Company, 3000 Hanover Street, Palo Alto, California 94304).

2. Adapted from A. Hax and N. Majluf, "The Corporate Strategic Planning Process," *Interfaces*, Vol. 14, No. 1, pp. 47–60, 1984. Used with permission. Copyright © 1984 by the Operations Research Society of America and The Institute of Management Sciences, 290 Westminster Street, Providence, Rhode Island 02903.

3. Adapted from an article by G. Adesso (1986).

Further Reading

Adesso, G. A., "Competitive Manufacturing in the Eighties," *Proceedings of the 1985 Annual Conference of the Association for Manufacturing Excellence*, J. B. Dilworth (ed.), Cincinnati, Ohio (Sept. 12-13th, 1985), 1986.

Fine, C., and A. C. Hax, "Manufacturing Strategy: A Methodology and an Illustration," *Interfaces*, Vol. 15, No. 6, pp. 28–46, 1985.

Fry, J. N., and J. P. Killing, *Strategic Analysis and Action*, Toronto: Prentice-Hall, 1986.

Hax, A. C., and N. S. Majluf, "The Use of the Growth-Share Matrix in Strategic Planning," *Interfaces*, Vol. 13, No. 1, pp. 46–60, 1983.

Hax, A. C., and N. S. Majluf, "The Use of the Industry Attractiveness-Business Strength Matrix in Strategic Planning," *Interfaces*, Vol. 13, No. 2, pp. 54–71, 1983.

Hax, A. C., and N. S. Majluf, "The Corporate Strategic Planning Process," *Interfaces*, Vol. 14, No. 1, pp. 47–60, 1984.

Kotha, A., and D. Orne, "Generic Manufacturing Strategies: A Conceptual Synthesis," *Strategic Management Journal*, Vol. 10, No. 1, pp. 211–231, 1989.

"Special Issue on Linking Strategy Formulation in Marketing and Operations: Empirical Research," edited by W. Berry, T. Hill, and C. McLaughlin, *Journal of Operations Management*, Vol. 10, No. 3, 1991.

Wheelwright, S., "Reflecting Corporate Strategy in Manufacturing Decisions," *Business Horizons*, Vol. 21, pp. 57–66, February 1978.

PART II

CHANGES AND TECHNIQUES FREQUENTLY USED IN MANUFACTURING STRATEGY

CHAPTER 9

Improvement Approaches in Manufacturing

With few exceptions, manufacturing companies all over the world are taking a close look at their production systems to make improvements. Five general improvement approaches are in widespread use today: total quality management (TQM), short cycle manufacturing (SCM), kaizen, agile manufacturing, and reengineering. There are three important ways in which these approaches differ (see Figure 9-1).

FIGURE 9–1

Manufacturing Improvement Approaches

Improvement Approach	Scope of Changes	Type and Frequency of Changes	Means for Identifying Problems
Total quality management (TQM)	Adjust levers*	Continuous, incremental changes	Quality
Short cycle manufacturing (SCM)	Adjust levers	Continuous, incremental changes	(Delivery) time
Kaizen	Adjust levers	Periodic, dramatic changes	Cost
Agile manufacturing	Adjust levers	Continuous, incremental changes	Flexibility
Reengineering	Change production system	Burst of breakthrough changes	Depends on the new production system

* No change is made to the type of production system used.

1. *Scope of changes.* Some approaches are used when major changes have to be made in a short time. Other approaches are used when the scope is smaller and the pace slower.
2. *Type and frequency of changes.* A steady stream of small, incremental changes may be sought; groups of dramatic changes may be sought at regular intervals (say, every four months); or a short burst of large, breakthrough changes may be sought.
3. *Means for identifying problems.* Each improvement approach focuses on a different manufacturing output as the means by which it identifies areas where improvements may be made.

All improvement approaches adjust the manufacturing levers to bring them into precisely the correct positions for the production system used. Once this is done, the improvement approaches continue to adjust the levers to raise the level of manufacturing capability. Most improvement approaches, including TQM and SCM, generate a steady stream of incremental changes. Approaches like kaizen seek larger changes at regular intervals (usually once every four months). Some improvement approaches, like reengineering, are used when the scope of the change is much larger, say, when an existing production system has to be changed to a new one. In this case, a short stream of large, breakthrough changes are needed to move all the manufacturing levers from their current positions to new positions appropriate for the new production system.

All improvement approaches require information on what adjustments other companies are making to their manufacturing levers. A technique called benchmarking is used to gather this information. Making improvements is not new. The theoretical underpinnings are learning and the product life cycle (see Chapter 11). What is new is the vigor with which the benefits of learning are sought. How benchmarking, TQM, SCM, kaizen, agile manufacturing, and reengineering are used in manufacturing strategy is discussed in the following sections.

BENCHMARKING

It is difficult to convince employees that changes are necessary. It is easier when strong images are available.

- How much better are we today than we were last year?
- How effective are we compared to our competitors? Compared to companies in other industries?
- How do product quality, setup times, cycle time for order processing, cycle time for product design, training budget, organization structure, and so on, compare to competitors in the same region? To the best competitors in the industry? To the best companies in other industries?
- Are we good enough?

Comparisons such as these provide motivation for change, but many organizations cannot make these comparisons because they lack the necessary information. Benchmarking is the technique used to gather this information. When an organization benchmarks, it looks for two types of information: 1) What are the values of performance measures (such as defect rate, delivery time, new products introduced per year, unit material cost, and participation rates) at other organizations? and 2) What practices are used by other organizations to achieve superior performance? These data are obtained from competitors in the same industry as well as from excellent companies in other industries. Xerox, for example, benchmarks more than 200 variables. They collect the following information for each variable: an industry average benchmark (the mean performance of companies in Xerox's industry), a competitive benchmark (the best performance in the industry), and a world class benchmark (the best performance in any industry).

Benchmark data can be obtained from periodicals and books; at professional meetings; and from discussions with professionals, suppliers, sales representatives, and other manufacturers. Obtaining information from organizations in the same industry is difficult because of the natural reluctance of companies to share information with their competitors. Obtaining information from a company in another industry is easier. Professionals working in companies that are not competitors are often happy to exchange information and ideas, and sometimes may be willing to share the expense of a benchmarking study. Small companies supplying products to larger companies can benchmark themselves against their customers.

Benchmarking is not easy. Gathering data is expensive and time-consuming. Even when data is available, it must be interpreted carefully. The data may come from products and processes that are slightly different, the environment may be different, and the data may come from an organization following a different strategy. A benchmarking study consists of the following steps:

1. Determine what information and what practices to benchmark.
2. Form a benchmarking team.
3. Identify sources of benchmark information.
4. Collect the information.
5. Analyze the information.
6. Use the information to motivate change.

Benchmarking is used in three elements of the manufacturing strategy—the competitive analysis, the level of manufacturing capability, and the manufacturing levers.

Competitive Analysis

A worksheet for benchmarking the six manufacturing outputs is shown in Figure 9-2. The worksheet is an expanded version of the competitive analysis element in the manufacturing strategy worksheet. Attributes for each manufacturing output are agreed on. Data are collected from companies in the same industry and from companies in other industries. These external data are compared with data from the company itself. Market qualifying and order winning outputs are then determined, and twelve-month and twenty-four-month targets are set.

Manufacturing Capability, Manufacturing Levers

A worksheet for benchmarking the best practices in use at other companies is shown in Figure 9-3. The worksheet is an expanded version of the manufacturing levers element in the manufacturing

Manufacturing Outputs		Company		Competitors from Same Industry								Competitors in Other Industries				Market Qualifying, Order Winning, or Not Important		
				Company A		Company B		Industry Average Competitor		Best Competitor		Company C		Company D		M, O	Targets	
	Measures	Current year	Last year	this year	last year	this year	last year	this year	last year	this year	last year	this year	last year	this year	last year		12 months	24 months
Cost	Unit cost Labor productivity Machine utilization Yield																	
Quality	Percent defective Rework costs Mean time between failures																	
Performance	Number of standard features Number of advanced features Product resale price																	
Delivery	Quoted delivery time Percentage on-time shipments Order entry cycle time Average lateness Number of expediters																	
Flexibility	Number of products in product line Number of options allowed Minimum order size Length of frozen schedule Average lot sizes																	
Innovativeness	Lead time to design new products Lead time to prepare customer drawings Number of engineering change orders per year Number of new products introduced each year																	

FIGURE 9 – 2

Worksheet for Benchmarking During the Competitive Analysis

Manufacturing Levers	Elements	Best Practices Used at Same/Other Companies in Same/Other Industries
		Output Affected: Delivery Cost Quality Performance Flexibility Innovativeness
		Market Qualifying/ Order Winning
Human Resources	Line operators Supervisors Maintenance Material handling Compensation Training, promotion	
Organization Structure and Controls	Performance measurement Capital budgeting Culture Departments Communication between departments	
Production Planning and Control	Order taking MRP system Shop floor control Inventories Lot sizes	
Sourcing	Quoted delivery time Percentage on-time shipments Make versus buy Supplier certification Supplier relations Purchasing	
Process Technology	Fabrication Assembly Waste treatment Mechanical technology Electronic technology New technology Automation	
Facilities	Size, focus Static, developing Location Age Shared resources	

FIGURE 9–3

Worksheet for Benchmarking Adjustments to Manufacturing Levers

strategy worksheet. Data on best practices are collected from companies in the same industry and from companies in other industries. Practices that can best provide the market qualifying and order winning outputs at the target levels are identified as adjustments that should be made to the manufacturing levers.

SITUATION 9.1

Benchmarking at Xerox[1]

WHEN XEROX started benchmarking in 1979, management's aim was to analyze performance measures such as unit costs, defect rates, and cycle times in its manufacturing operations. By the mid 1980s the benchmarking focus moved to analyzing the practices used by other organizations to achieve superior performance.

Anxious to improve its warehousing and distribution operations, Xerox started a benchmarking project in 1981. After studying the trade literature and talking to professionals in the field, Xerox discovered the L.L. Bean Co. of Freeport, Maine. L.L. Bean is a sporting goods retailer and mail-order house. What L.L. Bean and Xerox had in common was a warehousing and distribution system that handled a wide variety of products. When the Xerox benchmarking team visited L.L. Bean in February 1982, they found many differences in performance measures and practices. L.L. Bean was three times more efficient than Xerox on two important measures; the number of customer orders completed per day per employee, and the total number of items on these orders. The benchmarking team documented the practices used by L.L. Bean to achieve this outstanding performance. The team also visited a drug wholesale company, an electrical components manufacturer, and a catalog service bureau. Xerox incorporated the benchmarking information into new performance targets and practices to be implemented in its warehousing and distribution operations.

Improvement Approaches

The primary objective of any improvement program is to raise the level of capability of the manufacturing levers so that higher levels of the manufacturing outputs can be provided. In most companies improvement programs started as quality improvement programs for three reasons: 1) In the 1970s, the quality of Japanese products exceeded that of North American products in many industries, so it was natural to focus manufacturing improvement efforts on raising the level of quality; 2) not only were there many well-known techniques for identifying and solving the problems that caused poor quality, the techniques were organized into many effective quality programs; and 3) perhaps most important, experiences with the quality programs showed that improvement efforts focused on raising the level of quality also produced improvements in cost and delivery.

More recent improvement approaches have moved beyond quality to consider other manufacturing outputs, such as (delivery) time and flexibility, and combinations of outputs, as a better way to identify areas where improvements can be made. Short cycle manufacturing (SCM), which is also called time based competition, uses time as a measure of waste. Agile manufacturing focuses on flexibility. Kaizen is different from TQM, SCM, and agile manufacturing in that it looks for waste directly. It targets either one manufacturing output for improvement (and looks at each manufacturing lever for problems affecting that output), or it targets a particular lever for improvement. TQM, SCM, kaizen, and agile manufacturing are alternative improvement approaches. An organization looking for an improvement approach should select the one that is easiest for it to implement. Reengineering is not an alternative to TQM, SCM, kaizen, or agile manufacturing because, unlike these approaches, its goal is not to eliminate waste from a process. Its goal is to replace the process entirely. It is used in manufacturing strategy only when the existing production system is changed to a new system.

From the point of view of manufacturing strategy, all improvement approaches should increase the level of manufacturing capability of each lever and raise the levels of all the market qualifying and order winning outputs. Any implementation of an improvement approach that falls short of this—that is, one that increases only the level of the output it naturally focuses on (quality in the case of TQM, delivery for SCM, and flexibility for agile manufacturing)—is considered a failure.

Total Quality Management (TQM)

"Quality is customer satisfaction." That is how TQM defines quality. At Procter & Gamble, for example, quality is "the unyielding and continually improving effort by everyone to understand, meet, and exceed the expectations of customers."[2] Sometimes this definition causes confusion because it is much broader than the definition used in manufacturing strategy. However, the definitions are not inconsistent. In TQM, quality is a customer determination, not a manufacturing or marketing determination. To the customer, quality includes all the manufacturing and nonmanufacturing outputs. Consequently, meeting and exceeding customer expectations in TQM means providing outputs at market qualifying and order winning levels, and seeking to raise these levels over time.

Clarifying the TQM definition of quality starts with defining the word customer. A customer is anyone who receives a product. This includes external and internal customers. A product is the output of a process and includes goods and services. Customer satisfaction is achieved through providing market qualifying and order winning levels of the manufacturing and nonmanufacturing outputs. Employees are customers when they receive goods and services from other processes, they are producers when they convert these goods and services into products, and they are suppliers when they provide products to customers. Improving quality requires that employees inform their suppliers of their quality needs, improve their production process, and understand the quality needs of their customers. This is done by means of "features" and "freedom from deficiencies." Features are characteristics of the products received by customers. In TQM, a company learns what features its customers want and then tries to meet and exceed these expectations. Freedom from deficiencies focuses on the processes. Processes are brought under control, and actions are taken when a process strays out of control. The capability of each process is assessed and processes are improved.

When TQM is implemented in large companies, it involves numerous people and processes and huge quantities of information. Well-planned quality systems are needed. Three such systems are Florida Power and Light's TQM system, ISO 9000, and the quality system of the Malcolm Baldrige National Quality Award. The last two are examples of well-known, effective, and nonproprietary quality systems (see the references at the end of this chapter).

Situation 9.2

Total Quality Management at Florida Power and Light[3]

IN NOVEMBER 1989, Florida Power and Light (FPL) became the first company outside Japan to win the coveted Deming Prize, which recognizes outstanding achievement for quality management. Things had not always been so good. In the late 1970s FPL experienced problems similar to those faced by many manufacturers. Fuel costs were rising, inflation was soaring, large capital expenditures seemed inevitable, customers were demanding more reliable service at lower cost, and new small power producers were entering the market. After visits to Japan by company employees, the FPL Quality Improvement Program was developed. It consisted of three elements: quality improvement teams (started in 1982), policy deployment (started in 1984), and quality in daily work (started in 1986).

By 1988, more than 1,500 quality improvement teams consisting of 10,300 employees were at work (a 70 percent participation rate). Every team followed a rigid, seven-step quality improvement process regardless of the size or importance of the quality problem considered. Policy deployment was how FPL ensured that the quality objectives in its corporate strategic plan were translated into strategies and actions at all levels of the organization. Quality in daily work was the most difficult of the three elements. It was the application of TQM to each employee's job. Three tasks comprised quality in daily work. First, the work process was stabilized so that the quality of the output was in control. Then changes were made to improve the process. Finally, the standards were updated to reflect the changes.

Through 1987, the quality improvement road was bumpy. Despite many successes, some difficult problems persisted. The company needed an external crisis to inspire its employees to achieve a higher level of quality. FPL decided to create an artificial crisis by trying to win the Deming Prize. Peer pressure to excel quickly spread through the company. No division or department wanted to be the culprit that prevented FPL from winning the prize. The crisis stimulated everyone to give the extra effort required to change FPL from an organization that employed quality practices to one where quality improvement was a way of life.

Situation 9.3

ISO 9000's Total Quality System

IN 1986, the International Organization for Standardization (ISO) published the first edition of a document entitled "Guidelines for Third Party Assessment and Registration of a Supplier's Quality System." A year later, the ISO 9000 standards series was released. (Revisions followed in 1994.) Britain quickly released British Standards Institute BS5750 and the United States followed with the Q90 quality standard. Although there are some differences in terminology, the three standards are almost identical. ISO 9000 was a response to three developments that occurred in the 1980s:

1. With the expansion of free trade in Europe, North America, and other parts of the world, the need arose for an international quality standard that would assure countries that the quality of materials produced outside a country's borders was high enough that the materials could be imported and used.
2. Companies were developing far-flung international production and purchasing networks. A common quality standard was needed to ensure that material and components produced or purchased in different parts of the world met the same high-quality standards.

3. Suppliers were spending a great deal of time adjusting their quality systems to satisfy the differing requirements of their customers. Many suppliers wanted a single quality standard administered by an independent third party that would satisfy the quality requirements of all their customers.

The ISO 9000 Quality Standard series is a system in which a third party assesses and registers a company's quality system. It is organized as follows:

- 9000—Guidelines for selecting the quality system models in 9001, 9002 and 9003.
- 9001—A quality system model for detecting nonconformance in design/development, through production, installation, and servicing.
- 9002—A quality system model for detecting nonconformance in production, installation, and servicing.
- 9003—A quality system model for detecting nonconformance in the final inspection and test.
- 9004—A guideline for developing a quality management system.
- 1011—Guidelines for auditing quality systems.

In most industries, ISO 9000 is the minimum or market qualifying level of quality. Twenty areas make up the ISO 9001 quality system, which is the most complete system in the standard:

1. Management responsibility
2. Quality system
3. Contract review
4. Design control
5. Document and data control
6. Purchasing

7. Control of customer-supplied product
8. Production identification and traceability
9. Process control
10. Inspection and testing
11. Control of inspection, measuring, and test equipment
12. Inspection and test status
13. Control of nonconforming product
14. Corrective and preventive action
15. Handling, storage, packaging, preservation, and delivery
16. Control of quality records
17. Internal quality audits
18. Training
19. Servicing
20. Statistical techniques

Many of the sections in the ISO standard use terms like "shall," "where applicable," and "as required." For example, section 4.20 of ISO 9001 states, "Where applicable, the supplier shall establish procedures for identifying adequate statistical techniques required for verifying the acceptability of process capability and product characteristics." How to determine applicability and the techniques to do so are left for the user to decide. This flexibility is perceived by some to be a strength of the standard, by others, a weakness. It is a strength because it gives companies the freedom to choose what is best for them. Companies with high levels of manufacturing capability have little difficulty making these choices. Companies with low levels of capability are sometimes overwhelmed by the number of options available, and become frustrated when they cannot make choices quickly. Another source of frustration for some companies is the emphasis on documentation and record keeping. Again this is more likely to occur at companies with low levels of manufacturing capability than at those with higher levels. Companies with industry average and higher levels of capability will already have up-to-date documentation and efficient record keeping systems. The major benefit of seeking ISO 9000 certification is not the certification itself, but the improvements made in the 20 areas. They increase the level of manufacturing capability, which results in higher levels of all the manufacturing outputs.

Short Cycle Manufacturing (SCM)

Cycle time in short cycle manufacturing is the time from the receipt of a customer order until the customer accepts delivery of the product. The terms *cycle time* in SCM and *delivery time* in manufacturing strategy are equivalent. Cycle time is made up of five cycles (see Figure 9-4): the order processing cycle, the product

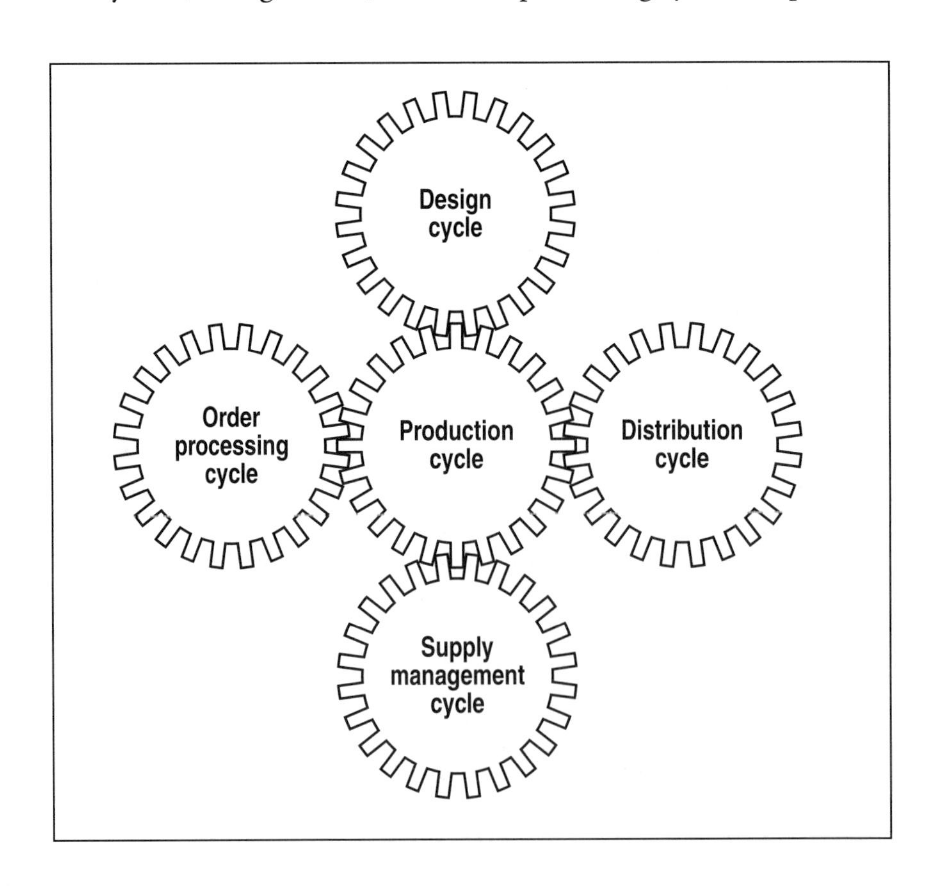

FIGURE 9–4

Components of Cycle Time

design cycle, the supply management cycle, the production cycle, and the distribution cycle. Numerous activities occur during each of these cycles (see Figure 9-5). Some activities are done sequentially and some concurrently, depending on which production system is used. For example, in a job shop production system where almost all products are custom designed, order processing and product design are done concurrently

SCM operates by reducing the amount of time available to accomplish all the activities in a cycle, for the purpose of eliminating any unnecessary activities. Sometimes activities can be eliminated outright, sometimes an indirect route must be taken.

Order processing cycle

- Respond to customer inquiries
- Create sales order
- Determine price
- Book production time in plant
- Check customer credit
- Determine promised delivery date

Design cycle

- Modify standard design to meet customer requirements
- Obtain customer approved drawings
- Design new products
- Conduct market research
- Analyze product technology
- Develop prototype
- Process planning

Supply management cycle

- Make versus buy decisions
- Identify, evaluate, and develop suppliers
- Negotiate terms
- Release orders for materials and components
- Release orders for tooling
- Set delivery dates

Production cycle

- Production planning and control
- Materials management
- Setups, processing
- Rework
- Maintenance

Distribution cycle

- Ship products to distribution centers
- Receive, count, inspect material
- Generate pick lists
- Pick material for customer orders
- Ship material to customer

FIGURE 9–5

Activities Making Up Cycle Timing

For example, poor quality may prevent activities such as inspection and rework from being eliminated. SCM tries to improve the quality of materials and procedures, and the reliability of equipment, so that inspection and rework can be eliminated.

SITUATION 9.4

Air Conditioner Manufacturer Implements Short Cycle Manufacturing[4]

IN 1991, Inter-City Products of Toronto, North America's second-largest manufacturer of home and commercial air conditioners, found itself with excess production capacity. Of its three plants, it would have to close either the Illinois or Ontario plant.

The Ontario plant had some disadvantages. Base wages and benefits were 30 percent higher than the Illinois plant. Local taxes were 400 percent higher. Freight costs to its major customers in the hot, southern U.S. states were 20 percent higher. However, the plant also had some advantages. In 1988, a major SCM program was started by Inter-City's senior vice president of operations and logistics. By the summer of 1991, the plant had reduced its cycle time, the time from receipt of a customer order to delivery of a finished air conditioner, from 22 days to 5 days. Work-in-process inventory was reduced from $21 million to $3 million, and the plant's capacity had doubled without increasing the size of the plant or the work force. Because of SCM, the plant could provide market qualifying and order winning levels of delivery, cost, quality, and flexibility.

In late 1991, the company made its decision. The Illinois plant would close. All commercial products would be produced in the Ontario plant, and all residential products would be produced in a plant in Tennessee.

SITUATION 9.5

Toyota Slashes Cycle Time in Order Processing and Distribution[5]

PRIOR TO 1982, Toyota's Japanese operations were divided into two separate companies: Toyota Motor Manufacturing and Toyota Motor Sales. The production cycle was less than two days at Toyota Manufacturing. However, the order processing cycle and the distribution cycle varied from 15 to 26 days at Toyota Sales. It took less than two days to manufacture a car, and between 15 and 26 days to take an order, transmit the order to a factory, schedule the order, and deliver the car to the customer.

The cost of sales and distribution at Toyota Sales was more than the cost of manufacturing the car at Toyota Manufacturing. In 1982, Toyota Motor Manufacturing and Toyota Motor Sales were merged. Within 18 months, all the Toyota Motor Sales directors retired. Their jobs were left vacant or filled by executives from Toyota Motor Manufacturing. SCM and other improvement programs were started. By 1987, the order processing cycle and the distribution cycle had been reduced to 6 days, and significant reductions in cost had been achieved.

Situation 9.6

Ford's Short Cycle Time at the River Rouge Plant[6]

HENRY FORD was one of the first to understand the importance of reducing time in manufacturing. The observations he made in Chapter 10, "The Meaning of Time," of his book *Today and Tomorrow* (written in 1926), on the improvements accomplished at the River Rouge plant are key principles in present-day approaches to cycle time reduction.

> Ordinarily, money put into raw materials or into finished stock is thought of as live money. It is money in the business, it is true, but having a stock of raw material or finished goods in excess of requirements is waste—which, like every other waste, turns up in high prices and low wages.
>
> The time element in manufacturing stretches from the moment the raw material is separated from the earth to the moment when the finished product is delivered to the ultimate consumer . . .
>
> If we were operating today [i.e., 1926] under the methods of 1921, we should have on hand raw materials to the value of about $120 million, and we should have unnecessarily in transit finished

> products to the value of about $50 million. That is, we should have an investment in raw material and finished goods of not far from $200 million. Instead of that, we have an average investment of only $50 million, or, to put it another way, our inventory, raw and finished, is less than it was when our production was only half as great.
>
> The extension of our business since 1921 has been very great, yet, in a way, all this great expansion has been paid for out of money which, under our old methods, would have lain idle in piles of iron, steel, coal, or in finished automobiles stored in warehouses. We do not own or use a single warehouse!
>
> How we do this will be explained later in this chapter, but the point now is to direct thought to the time factor in service. . . . [pp. 112–113]
>
> Our production cycle is about 81 hours from the mine to the finished machine in the freight car, or 3 days and 9 hours instead of the 14 days which we used to think was record breaking. [p. 118]

Kaizen

Kaizen means continuous, incremental improvements. It is a hands-on, do-it-now approach that incorporates a few dramatic improvements every few months. A simple implementation consists of the following four steps (see Chese and Tanner 1993).

Step 1: Organize

Every three or four months, management organizes three to six kaizen teams of eight to ten members each. New teams are formed for each session, with team members coming from all parts of the organization. Areas where improvements can be made (called "kaizen points") are identified. Specific, ambitious goals are set for each kaizen point. Employees are assured that they will not be affected adversely by kaizen. Teams are charged with making improvements at the kaizen points. There are few restrictions on the changes the team is permitted to make. It can eliminate operators, reduce inventories, move equipment, compress layouts, modify tooling, revise procedures, and so on. Kaizen sessions last 2½ or 3½ days. During this time, support personnel such as plant trades, engineering, tooling, and material handling are on "kaizen alert" to provide immediate assistance to the teams.

Step 2: Learn

Each kaizen session starts with a half day for learning and developing action plans. It is important to guard against prolonged discussion that might lead to "pencil kaizening" (that is, "paralysis through analysis").

Step 3: Hit the Floor Running

After the half day of learning, the team is dispatched to the plant floor. Kaizen team members, attired in bright shirts and caps, are now visible all over the plant. Accurate observations and records are mandatory. Layouts are changed, procedures are revised, tooling is changed, operators are moved, and documentation is prepared. Every two or three hours, the team reassembles to review progress and then returns to the floor.

Step 4: Boast

Each kaizen session ends with team presentations. Pride in accomplishment and admiration within and across teams allows for good-natured boasting and bragging. After the last team has finished, the session leader recaps the accomplishments, praises the teams, recounts the principles of kaizen, and rewards each participant.

Agile Manufacturing

Agile manufacturing is the application of new, affordable, flexible automation to a production system for the purpose of increasing flexibility. Agile manufacturing enables any production system to produce a more varied product mix in lower volumes than was possible before. For example, an agile operator-paced line flow production system can produce what in the past could be produced only on a batch flow production system (see Situation 9.7).

Agile manufacturing is a shift to the left of the production systems in the PV–LF matrix (see Figure 9-6). Implementing agile manufacturing is not easy. Flexible automation, concurrent engineering, and information technology are the three foundations on which it is built. It requires significant increases in the levels of capability of the organization structure and control lever, the process technology lever, and (for most manufacturers) modest increases in the levels of capability of the other four manufacturing levers. Implementing it raises the level of capability of the production system, which leads to increases in the levels of all the manufacturing outputs.

FIGURE 9–6

Agile Production Systems

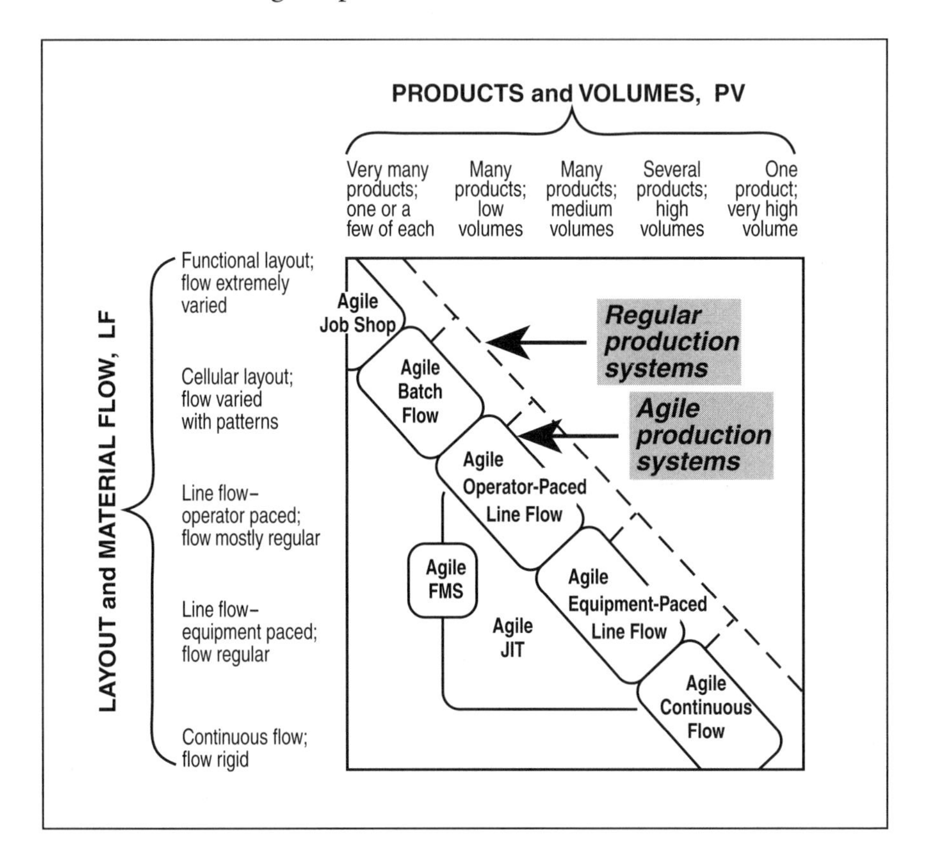

SITUATION 9.7

Agile Manufacturing at a Bicycle Manufacturer[7]

NBI was one of Japan's largest bicycle manufacturers. It used three different production systems to supply about 9 percent of the Japanese market of 8 million bicycles a year. Standard design bicycles, which sold for about $370, were mass-produced on equipment-paced line flow production systems. Custom bicycles were produced in a job shop production system. Built by skilled craftspeople to each customer's individual body measurements, riding style, and preference for components, accessories, and finishing, these custom bicycles took two months to build and sold for about $3,000. In 1987, NBI began producing "personalized semicustom bicycles" on an agile operator-paced line flow production system. It took eight to ten days to produce a bicycle, and 50 to 60 were produced every day. Each semicustom bicycle, which sold for about $1,300, was treated as a package of options. Options included 18 models, 15 frame sizes, 18 handlebar styles, 6 pedals, 4 tires, and more than 200 finishes. This represented a large number of different semicustom bicycles, but it was a small fraction of the number of possible custom bicycles.

The agile operator-paced line flow production system worked as follows. Customer orders were taken at bicycle shops all over Japan. The customer's body measurements were taken and entered, with his or her preferences, on a detailed order form. The form was faxed to the factory, where operators immediately entered the information into a computer. The computer generated a CAD/CAM drawing of the bicycle, assigned a bar-code number for all components, and released an order to the factory. At the factory, each bicycle was assigned to a veteran craftsperson who was responsible for overseeing the process from beginning to end. Almost all fabrication operations such as cutting, brazing, gluing, checking dimensions, cleaning, and painting were done by robots. Finer, more detailed operations, such as fitting, assembly, and finishing, were done by the craftsperson. Once complete, the personalized bicycle was packed and shipped to the shop where the original order was placed. The shop completed the assembly and presented the bicycle to the customer.

Whereas NBI used an agile operator-paced line flow production system, the mix and volume of products and the market qualifying and order winning outputs were such that the regular batch flow and JIT production systems could also have been used.

SITUATION 9.8

Motorola's Agile Equipment-Paced Line Flow Production System

IN THIS EXAMPLE, an agile equipment-paced line flow production system was used to produce a mix and volume of products that would have been produced on an operator-paced line flow production system before the availability of flexible automation. In 1987, after 18 months of design work, Motorola started a state-of-the-art equipment-paced line flow production system to manufacture its Bravo pocket pager. The production system produced a small family of pagers, each using the same printed circuit board and working on a single frequency band, with high levels of cost, quality, and delivery manufacturing outputs. In the years that followed, improvements were made and flexible automation was added. By 1989, more than one board could be processed at the same time. In 1990, products with different frequency bands could be produced. In 1991, different colors could be provided. Then, in 1993, a new agile equipment-paced line flow production system was organized to manufacture products with different boards and different bands in single-unit lots. Concurrent with the development of the agile production system was a complete redesign of the pagers. Agility was achieved through new flexible automation, carefully designed new products, and the many improvements made to the previous production systems.

Three factors account for the popularity of agile manufacturing:

1. Agile manufacturing makes it possible for companies to avoid using the difficult JIT and expensive FMS production systems. Agile operator- and equipment-paced line flow production systems produce the same mix and volume of products as JIT and FMS. They are much easier to manage, and the levels at which the manufacturing outputs are provided are similar.
2. Many companies are anxious to gain some experience with the hardware and software that make up flexible automation in agile manufacturing because they believe it could become part of their future core process technology.

3. Implementing agile manufacturing involves purchasing new equipment, working with external suppliers to develop flexible automation, and installing new technology. These are tasks that manufacturing engineers enjoy.

Reengineering

Reengineering was a response to two business problems of the late 1980s. First, many companies realized that cost cutting and incremental improvement approaches like TQM, SCM, and kaizen could not produce the huge improvements their companies needed to survive. More fundamental change was needed. Business processes had to be completely redesigned. Second, many companies could not obtain a return from their huge investments in information technology. Putting the two problems, business process redesign and information technology together, produced reengineering. In reengineering, existing business processes and the underlying rules and assumptions are discarded in favor of new processes that make extensive use of information technology. Information technology is not used to automate existing processes; it is used to enable new ones.

Process redesign is well known in manufacturing. What is different about reengineering is the linking of process redesign with information technology, the application of process redesign to all business processes, and the goal of achieving a major improvement on the bottom line. Because of the need to achieve major bottom-line results, only those business processes whose scopes are broad and deep are targeted for reengineering. These processes require a lot of time and resources to reengineer. A team of the company's most skilled employees, committed full-time and almost always aided by external consultants, can take about one year to envision, design, prototype, test, and prove the benefits of a typical reengineered process. Another year or two may be necessary to build the redesigned process and implement it throughout the organization.

Reengineering is a useful improvement approach when manufacturing is changing from one production system to another. Manufacturing must break away from the assumptions and rules of the current production system and adopt new ones appropriate for the new production system. It requires major changes to all the manufacturing levers. After reengineering changes the production system, TQM can be used to stabilize and control the system, and TQM, SCM, and kaizen can be used to scrutinize it for incremental improvement opportunities. Six steps make up a reengineering project.

Step 1: *Develop objectives.*

Step 2: *Identify the processes to be reengineered.* Each of the five cycles that constitute cycle time (see Figure 9-4 on page 172) is a business process. Business processes have three important characteristics. They have customers, they cross organizational boundaries, and they are large enough to have a major effect on the business's bottom line. Most organizations could benefit from a redesign of all business processes. However, the amount of effort involved creates practical limitations. Most organizations have some sense of which business processes are most critical to their success and are most in need of change. Few organizations can support more than one redesign a year.

Step 3: *Understand and measure the existing process.* It is important to understand and measure processes before attempting to redesign them. Problems must be understood so that they are not repeated. Measurements should be taken so that improvements can be tracked.

Step 4: *Identify information technology capabilities and other process enablers.* Benchmarking is used to find process enablers such as information technology that other firms use to make their processes work more efficiently. Visioning is used to generate ideas for new processes by identifying and discarding old assumptions.

Step 5: *Design and build a prototype of the new process.*

Step 6: *Implement the new process.*

STANDARDS

Standards are an important part of any improvement program. When improvements are made, they are recorded and become the accepted way in which operations will be done in the future. Updating the standards each time an improvement is made prevents the organization from slipping back and losing the benefits gained. Standards are used as a base from which further improvements can be made.

Henry Ford was a strong believer in standards. In 1926, he wrote:

> To standardize a method is to choose, out of the many methods, the best one, and use it . . . What is the best way to do a thing? It is the sum of all the good ways we have discovered up to the present. It therefore becomes the standard . . . Today's standardization, instead of being a barricade against improvement, is the necessary foundation on which tomorrow's improvement will be based . . . If you think of standardization as the best that you know today, but which is to be improved tomorrow—you get somewhere. But if you think of standards as confining, then progress stops.[8]

Summary

TQM, SCM, kaizen, and agile manufacturing are improvement approaches that increase manufacturing capability and raise the level of the manufacturing outputs. Reengineering is used when the production system must be changed. Benchmarking provides the information necessary to implement the improvement approaches—information on performance and best practices. It is important to guard against losing improvements after they are made. Standards capture improvements and provide the starting points for subsequent improvements.

Notes

1. See Tucker, Zivan, and Camp 1987 and Camp 1990.
2. Procter & Gamble quality definition quoted in Lindsay and Evans 1993, p. 16.
3. Adapted from "Building a Quality Improvement Program at Florida Power and Light," *Target*, Vol. 4, No. 3, pp. 4–12, Fall 1988; and "For Florida Power and Light After the Deming Prize: The 'Music' Builds . . . And Builds . . . And Builds," *Target*, Vol. 6, No. 2, pp. 10–21, Summer 1990.

4. *The Toronto Star*, p. B8, Friday, November 1, 1991. See also P. Northey and N. Southway, *Cycle Time Management*, Portland, OR: Productivity Press, 1992, for more about ICP's cycle time reduction.

5. G. Stalk, "Time—The next source of competitive advantage," *Harvard Business Review*, pp. 41–51, July–August 1988.

6. Henry Ford, with Samuel Crowther, *Today and Tomorrow*, reprinted by Productivity Press, Portland, Oregon, 1988.

7. Adapted from T. E. Bell, "Bicycles on a Personalized Basis," *IEEE Spectrum*, pp. 32–35, September 1993.

8. Henry Ford, with Samuel Crowther, p. 82, ibid.

Further Reading

Benchmarking

Camp, R. C., "Competitive Benchmarking: Xerox's Powerful Quality Tool," *Making Total Quality Happen*, New York, Conference Board, 1990.

Tucker, F., S. Zivan, and R. Camp, "How to Measure Yourself against the Best," *Harvard Business Review*, p. 8, January–February 1987.

Total Quality Management

Evans, J., and W. Lindsay, *The Management and Control of Quality*, St. Paul: West Publishing, 1993.

Juran, J. M., and F. M. Gryna, *Quality Planning and Analysis*, New York: McGraw-Hill, 1993.

Malcolm Baldrige National Quality Award, National Institute of Standards and Technology, Route 270 and Quince Orchard Road, Administration Building, Room A537, Gaithersburg, MD 20899-0001.

Short Cycle Manufacturing

Ford, H., with S. Crowther, *Today and Tommorrow*, Garden City, NY: Doubleday, Page & Co., 1926 (reprinted by Productivity Press, Portland, Oregon, 1988).

Northey, P., and N. Southway, *Cycle Time Management: The Fast Track to Time-Based Productivity Improvement*, Portland, OR: Productivity Press, 1993.

Stalk, G., "Time—The Next Source of Competitive Advantage," *Harvard Business Review*, pp. 41–51, July–August 1988.

Kaizen

Chese, R., and C. Tanner, "Critikon Declares War on Waste, Launches Kaizen Drive," *Target*, Vol. 9, No. 4, pp. 12–22, July/August 1993.

Imai, M., *Kaizen*, New York: Random House, 1986.

Reengineering

Davenport, T. H., and J. E. Short, "The New Industrial Engineering: Information Technology and Business Process Redesign," *Sloan Management Review*, pp. 11–27, Summer 1990.

Hall, G., J. Rosenthal, and J. Wake, "How to Make Reengineering Really Work," *Harvard Business Review*, pp. 119–131, November–December 1993.

Hammer, M., and J. Champy, *Reengineering the Corporation: A Manifesto for Business Revolution*, New York: Harper Collins, 1993.

CHAPTER 10

Focus, Soft Technologies, Hard Technologies

Regardless of the improvement approach used (see Chapter 9), there is a sequence in which types of improvements should be made. First, manufacturing is focused. Then, soft technologies are used to improve the focused operations. Finally, hard technologies are added. Many attempts to improve manufacturing fail because one of these steps—focus, soft technologies, hard technologies—was missed or the steps were done in the wrong order.

Focus

Many organizations try to do too many tasks in the same plant. The result is a plant that does not do anything particularly well. This is tolerable when competitors behave in the same way. However, when competitors organize their plants into small plants-within-a-plant (PWPs), where each PWP focuses on a limited number of products or activities, the large unfocused plants can no longer compete. Manufacturing is focused by organizing PWPs and selecting the best production system—that is, the system most capable of providing the market qualifying and order winning outputs at the target levels—for each. The benefits of focused manufacturing include the ability to use the best production system, effective cooperation between functional areas, improved flow of materials, improved reaction to problems, closer ties to the market, more opportunities to improve, and better cost accounting.

There are situations where focus is not beneficial. For example, it may not be appropriate to focus when the products are seasonal or are in the decline stage of their product life cycle (see Chapter 11).

SITUATION 10.1

Furniture Company Focuses Manufacturing[1]

FOR 12 YEARS after Fernand Fontaine founded the company, Dutailer made rocking chairs and living-room and bedroom sets. In 1988, the Quebec company eliminated the living-room and bedroom sets, and began concentrating solely on rocking chairs. "That was the turning point. Before that, we were like everybody else in the Canadian marketplace, trying to be all things to all people."

The company produced its Glider Rocker in 35 models, with 12 different finishes from plain, varnished pine to each season's most fashionably colored lacquers. Customers could choose among 60 chair coverings, from printed fabrics to sophisticated Italian leathers. Finishes and fabrics were changed twice a year in response to changes in customer tastes.

Specialization worked for Dutailer. In 1991, 550 people were employed in four plants, up from fewer than 200 in 1988. Seventy-five percent of customers were in the United States (where Sears is a major customer) and Europe.

SITUATION 10.2

Focusing Manufacturing at AT&T

THE AT&T Dallas Works plant in Mesquite, Texas, was the largest manufacturer of power systems in the world. In 1989, the plant was completely rearranged and modernized. Four focused factories—board mounted power, off-line switchers, energy systems, and converters and mature products—were organized. The cost of the multimillion dollar project was recovered in less than a year. Delivery time reductions of up to 60 percent were achieved. Material costs were reduced and quality was improved. Even more focus was achieved by moving product management and marketing jobs from the AT&T Bell Lab in New Jersey to the plant in Mesquite. This made each focused factory the center of all activity—from design to manufacturing to marketing—for its product family.

SITUATION 10.3

Automotive Parts Producer Focuses Manufacturing[2]

THE XYZ plant produced parts for the automotive industry. During the 1970s growth came through the addition of new products. This helped absorb fixed overhead costs; however, the increase in the number of products made planning and control of manufacturing activities more difficult, even with the numerous improvements made to the computerized information systems. In the 1980s worldwide competition arrived, and customer requirements for cost, quality, delivery, performance, flexibility, and innovativeness began to increase at an alarming rate.

The plant considered three possible solutions:

1. Become a computer-integrated organization in which sophisticated computer programs would direct and control all manufacturing activities from order entry, through fabrication and assembly, to shipping.
2. Build smaller, independent plants with 100 to 200 employees.
3. Organize small plants-within-the-plant, each with 50 employees, to produce similar products or perform similar activities.

The first alternative was rejected because of the earlier disappointment with computerized information systems. The second alternative was rejected because of the high cost of building new facilities. That left the last alternative. Small PWPs were organized. Each was run by a small management team, and each was responsible for its own quality, scheduling, costs, and employee attendance. Each

PWP developed budgets, forecasts, and long-range plans. When possible, each PWP had direct contact with customers. Competition among PWPs was encouraged, but top management checked that the self-interests of the PWPs supported rather than conflicted with the overall goals of the organization.

SITUATION 10.4

Yacht Manufacturer Organizes Plants-within-the-Plant[3]

PROBLEM

GRE Yachts built custom design yachts. The company enjoyed a reputation for excellent design and workmanship. To capitalize on its reputation and secure a piece of the growing market for standard, fixed design yachts, GRE designed and started to manufacture a standard yacht. The yacht was aimed at the high end of the standard yacht market and had many of the features found on the company's custom design yachts. However, even this end of the market was more price sensitive and less conscious of performance than were GRE's custom design customers.

All of the company's yachts were manufactured in the same plant and shared the same equipment and skilled labor force. Custom design yachts were given priority in production because their profit margins were higher. Over the past year, as the standard yacht continued its steady increase in sales, costs and deliveries of all yachts began to slide. Many standard yachts were strewn around the yard in various stages of completion.

SOLUTION

The solution to GRE's problems is clear once a manufacturing strategy is developed. Recall from Chapter 6 the three-step procedure for developing manufacturing strategy: Step 1: Where is GRE? Step 2: Where does GRE want to be? Step 3: How does GRE get from where it is to where it wants to be?

Step 1: Where Is GRE?

GRE manufactured many products (different custom design yachts as well as the standard yacht) in volumes ranging from "one of" to "low volumes." The plant layout was a combination of functional and cellular. The material flow was varied with patterns. The production system was a batch flow (see Figures 10-1 and 10-2). GRE's level of manufacturing capability was industry average for all manufacturing levers except production planning and control, which was lower because of the current scheduling problems and high levels of work-in-process inventory.

Step 2: Where Does GRE Want to Be?

Custom design yachts. In its custom design yacht business, GRE provided high levels of performance (fast, attractive, comfortable yachts), quality (excellent workmanship), and innovativeness and flexibility (the ability to produce a yacht that satisfied a customer's unique requirements). These were the market qualifying outputs. Innovativeness was the order winning output. This meant that designers, suppliers, and manufacturing had to work closely together so that new designs could be completed and manufactured quickly and easily. Although the designs would be new, they might not always include the newest, most advanced features in the industry, which would be the case if performance was the order winning output (see Figure 10-1).

Standard design yachts. The standard design yacht is a single product produced in medium to high volumes (relative to the production volume of a custom design yacht). The market qualifying outputs were cost, quality, delivery, and performance. Flexibility and innovativeness were not as important in the standard yacht business. Quality was the order winning output, and took advantage of GRE's reputation in the custom design yacht business (see Figure 10-2). The existing batch flow production system is not

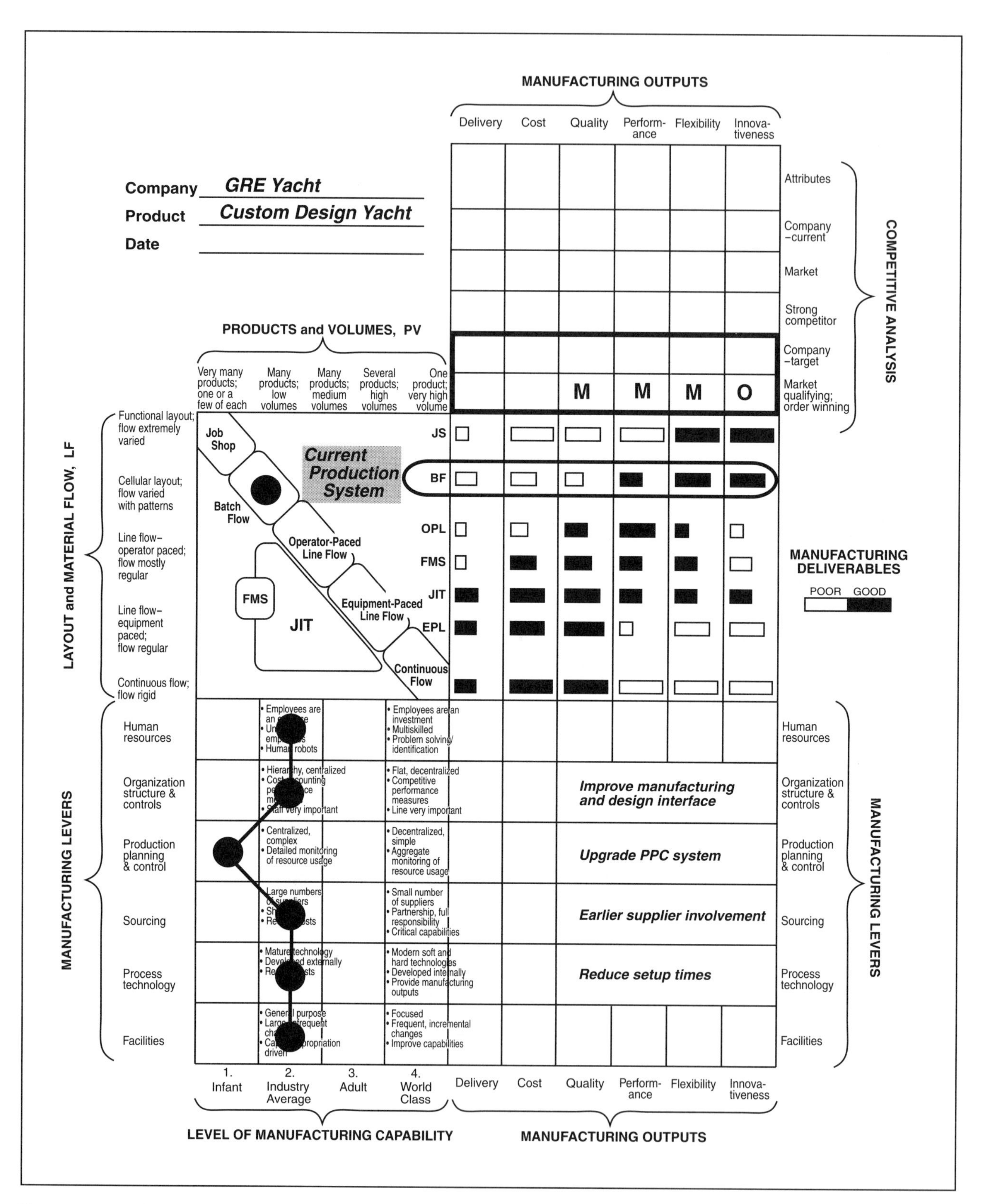

FIGURE 10–1

Manufacturing Strategy Worksheet for the Custom Design Yacht

appropriate for this product mix and volume, or for the market qualifying and order winning outputs. The operator- or equipment-paced line flow production systems would be more appropriate.

Step 3: How Does GRE Get from Where It Is to Where It Wants to Be?

Custom design yachts. Figure 10-1 shows that the existing batch flow production system can provide the market qualifying and order winning outputs, provided all the manufacturing levers are set in the proper positions and the level of manufacturing capability is sufficiently high. Several adjustments to the levers are required. The level of capability of the production planning and control lever must be raised to the same level as the other levers. This can be done by implementing an MRP system for make-to-order manufacturers. The relationship between design and manufacturing should be strengthened (because innovativeness is the order winning output) by starting concurrent engineering and design for manufacturing programs. The capability of the process technology lever can be raised by starting a setup time reduction program. The capability of the entire production system can be raised by adopting the kaizen improvement approach.

Standard design yachts. An operator- or equipment-paced line flow production system is required for the standard design yacht (see Figure 10-2). This is a different production system from the existing batch flow system, and so a separate PWP must be created. Each manufacturing lever must be adjusted so that it is in an appropriate position for the new production system. Adjustments include 1) using less skilled employees; 2) assigning engineering staff to design, organize, and support the line, to prepare quality control procedures, and so on; 3) developing close ties with suppliers who will supply larger volumes of standard parts (suppliers will be selected on the basis of cost, quality, and delivery, and more components will be outsourced); 4) implementing a line flow scheduling

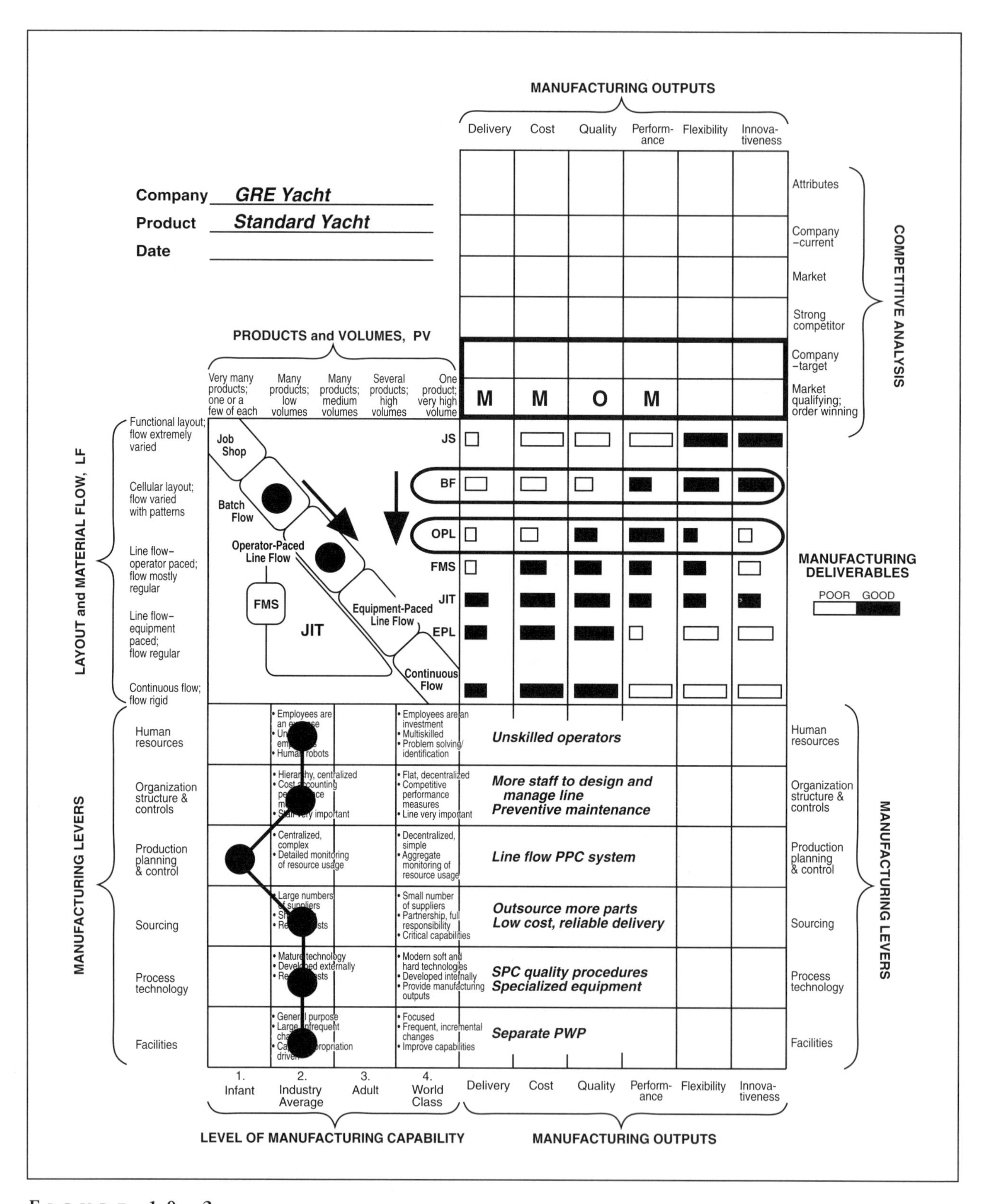

FIGURE 10–2
Manufacturing Strategy Worksheet for the Standard Yacht

system and a standard MRP system for controlling materials and activities; and 5) using specialized machines and tooling. Because quality is the order winning output, the TQM improvement appoach would be appropriate for this production system.

Because the level of manufacturing capability is only industry average (with a batch flow rather than a line flow production system), it will not be easy to make all these changes. Consequently, GRE should set up a small operator-paced line flow production system in a separate PWP. They should begin by producing a small number of the standard design yacht. As experience with the new production system is gained, the volume can be increased.

SITUATION 10.5

Focused Manufacturing at Cummins*

THE CUMMINS ENGINE COMPANY manufactured diesel engines. For years its NT product line was stable, manufacturing operated splendidly, and the company held more than 50 percent of the market. All this changed in the mid-1970s when the U.S. government introduced emissions regulations, and competition with Komatsu and Caterpillar escalated. Customers demanded faster deliveries, more flexibility, and lower prices. Cummins was forced to launch seven generations of the NT engine and three new engine families. The company cut prices on new products from 20 percent to 40 percent just to retain market share.

In 1986, the Cummins catalog offered over 100,000 parts. Although design engineers were becoming more sensitive to the difficulties of manufacturing so many different parts, the number of parts was still rising. For example, one of the newest engine families had 86 different flywheels, 49 flywheel housing options, and 17 starter motors with 12 possible mountings, all of which

* Reprinted by permission of the *Harvard Business Review*. An excerpt from "Cummins Engine Flexes Its Factory" by Ravi Venkatesan, March/April 1990.

could be assembled in approximately 1,200 different ways. Manufacturing also had to handle design changes, produce replacement parts for old engines, and react to cyclical demands. Between 1980 and 1985, most of the company's factories, frustrated by high costs, long delivery times, and high work-in-process inventory, made the transition to focused, or what Cummins called "cellular," manufacturing. One plant, for example, was reorganized into 15 cells. Each cell manufactured a small family of related parts. There was a water pump cell, a flywheel cell, a manifold cell, and so on. It soon became apparent however, that cellular manufacturing techniques could not meet the challenge of a greatly proliferated product line. The problem was setup time. Cells that were organized to produce 7 parts efficiently in batches of 5,000 were now expected to produce 14 parts in batches of 500. With so many parts (many with low volumes), it became impossible to change tools and fixtures quickly enough. The low volume parts started to choke the cells.

In 1987, the company commissioned a small team to develop insights into the problem of reorganizing its plants for the proliferated product line. The team determined that aggressive setup reduction was required and, equally important, that Cummins's factories should be focused by volume as well as by product. The team was able to discern four classes of products and five classes of volume. The product classes were skeletal parts (such as the block, head, and connecting rod), engine subsystem parts (such as the water pump and the lubricating pump), application parts (such as the exhaust manifold and the flywheel housing), and performance parts (such as turbochargers and compression brakes). The volume classes were defined by three parameters: volume (high, medium, or low), predictability of demand (predictable or unpredictable), and stability of design (relatively stable or rapidly evolving). Parts with the same product class and volume class were grouped into a family. All parts in a family were produced by the same production system. The only exception was that, when a part had setup times that were vastly different from the other parts in the family, the part would be produced on a different production system.

Four different production systems were used to manufacture all the families—the job shop (a traditional machine shop consisting mostly of CNC machines), batch flow, FMS, and equipment-paced line flow. Figure 10-3 and the discussion that follows describes how this was done in one Cummins plant.

Blocks

Blocks A and B are in the same product class. Both are in the same volume class because either the volume is high (block A) or the setup time is short (block B). Consequently, both are produced by the same production system, in this case, an equipment-paced line flow system because the total volume is high and the design is stable.

FIGURE 10–3

Selecting Production Systems for Part Families at Cummins

Part	Product Class	Annual Production Volume	Delivery Frequency	Design Stability	Time to Change Over	Production System
Block A	Skeletal	75,000	Daily	Stable	Changeover from A to B takes 30 minutes.	EP line flow
Block B		5,000	Monthly	Stable		Same EP line flow
Water pump A	Engine subsystem	75,000	Daily	Two-year design life	A and B have common features. Changeover from A to B takes 8 hours. C is a very different pump. Changeover from A to C takes 2 days.	FMS
Water pump B		2,000	Unpredictable	Stable		Same FMS
Water pump C		5,000	Daily	Stable		Batch flow
Manifold A	Application	75,000	Daily	Stable	Changeover from A to B takes 8 hours on a boring machine. Changeover from A to C takes 2 days.	EP line flow
Manifold B		5,000	Monthly	Stable		Same EP line flow
Manifold C		20	Unpredictable	Stable		Job shop

Source: R. Venkatesan 1990.

Water Pumps

Water pumps A, B, and C share the same product class but are in two volume classes. The volume of pump A is high, but it cannot be produced by an equipment-paced line flow production system. Such a system would require an investment in specialized equipment, which could not be recouped because the pump has a design life of only two years. In this plant, high volume parts with rapidly evolving designs were manufactured in flexible machining cells (along with prototypes and new products). Pump B, which shares many design features with pump A, has a low, unpredictable demand. The setup time for pump B is eight hours on traditional equipment and only one hour on FMS. Consequently, it is assigned to the same FMS as pump A. The flexibility of FMS is wasted on pump C. This product has a stable design, medium volume, and predictable demand. The most economical way to produce it is to use conventional equipment in a conventional cell (the batch flow production system).

Manifolds

Manifolds A, B, and C are in the same product class but are in different volume classes. Manifold A is produced on an equipment-paced line flow production system because of its high volume, stable design, and predictable demand. It would be difficult to manufacture manifold B on the same line because it takes eight hours to set up a conventional boring machine. The setup time on a CNC boring machine would be minimal. A practical solution is to use a CNC boring machine rather than a conventional one in the line producing manifold A. This flexible machine will slow down the line's speed, but it would permit manifold A and B to be produced on the same line. Manifold C, with its lengthy setup time and unpredictable demand, would be too disruptive to produce on the line producing A and B. Thus, it is assigned to the traditional machine shop (job shop production system).

Flywheel Housings

Another product class was flywheel housings (not shown in Figure 10-3). A total of 160 flywheel parts were divided into a high-volume (11 parts representing 85 percent of the volume) and a low-volume family. The high-volume family was produced on a dedicated line in an equipment-paced line flow production system. The 149 parts in the low-volume family were produced on a newly acquired flexible machining cell (FMS production system). Significant improvements in cost, quality, and delivery were achieved from focusing production of these products (see Figure 10-4).

Performance Measure	Before Focusing	After Focusing
Labor efficiency	42%	96%
Hours of output per operator	3.6	7.7
Number of operators	43	18
Number of salaried employees	6	2
Throughput time	1.8 days	40 minutes
Internal distance traveled	650 feet	256 feet
Inventory reduction		
• Raw material	6 days	1 day
• Work-in-process	1.4 days	1 hour
• Finished goods	4 days	1 day
On-time delivery	30%	100%
Product cost	--	43% reduction
Scrap	--	99% reduction
Full-time rework employees	3	0
Full-time inspectors	3	Audit only

Source: R. Venkatesan 1990.

FIGURE 10–4

Results of Focusing Production of Flywheel Housings

SOFT TECHNOLOGIES

The many new techniques and technologies of the last 15 years can be divided into two groups; soft technologies and hard technologies. Soft technologies improve manufacturing infrastructure. They are systems and people oriented, inexpensive to acquire, and difficult to implement (see Figure 10-5). Certain soft technologies work more efficiently with some production systems than with others. For example, total productive maintenance, an approach where operators perform routine maintenance tasks on the equipment they operate, is as effective in a job shop production

FIGURE 10–5

Some Soft Technologies

Concurrent engineering

A program wherein design engineering and process engineering work together from the start of product design through pilot production until the product is being produced on a routine basis. The objective is to optimize product design and process design.

Housekeeping

A program for organizing the work area and keeping it tidy.

Problem-solving techniques

Techniques for identifying and prioritizing problems, and determining appropriate solutions.

Process capability

A measure of the amount of variation in a process relative to the tolerances for the products being produced.

Pull production control system

A system where production at an upstream operation is triggered by consumption of a product at a downstream operation.

Setup time reduction

A program to reduce the amount of time required to set up, or change over, a manufacturing operation so that production of a new product may begin.

Small lot production

A program to facilitate the economical production of small batches of products.

Standardization

A program to reduce the amount of variety in products and processes.

Statistical process control

The use of statistical techniques to monitor and control the variation in a process.

Supply line management

A program of managing relationships with suppliers that includes supplier certification, monitoring and improving quality, delivery, and cost, and facilitating supplier participation in product design.

Team approaches

An approach wherein groups of employees work together to achieve common goals.

Total productive maintenance

A program wherein production operators participate in the maintenance of their equipment.

system as it is in an equipment-paced line flow production system. However, it does not work well in an FMS or a continuous-flow production system because the equipment is complex and the operators are relatively unskilled. Only those soft technologies appropriate for the production system that help the system provide higher levels of the market qualifying and order winning outputs should be implemented. To illustrate the character of soft technologies, we consider two popular ones: process capability and concurrent engineering.

Process Capability

NML had supplied parts from one of its North American plants to its Japanese customer for many years. There had never been any problems with cost, quality, or delivery. However, the customer had sent a purchasing team to NML to collect information for use in deciding whether to continue the relationship. Among other things, the team wanted to determine the capability of the manufacturing process, so it collected data to calculate process capability indices (C_p).

The critical quality attribute for one important part was the width of a slot. The specification was 400 ± 10. Two hundred parts were examined over a one-week period (see Figure 10-6). No slot had a width of more than 406 or less than 393, so there were no defects (parts with slot widths outside the specifications). The average or mean width was 398.62, and the standard deviation was $\sigma = 2.17$. Standard deviation (which is also called sigma, σ) is a measure of variability—in this case, the variability in the widths of the slots.

The process capability index, C_p, is defined as:

$$C_p = \frac{\text{specification width}}{\text{6 standard deviations}}$$

So $C_p = (10 + 10)/(6 \times 2.17) = 1.54$.

A more accurate measure takes account of the difference between the target and the mean of the process.

$$C_{pk} = C_p \times \left(1 - \frac{k}{\text{specification width / 2}}\right)$$

where $k = |\text{ target } - \text{ mean }|$.

FIGURE 10-6

Values of Quality Attribute

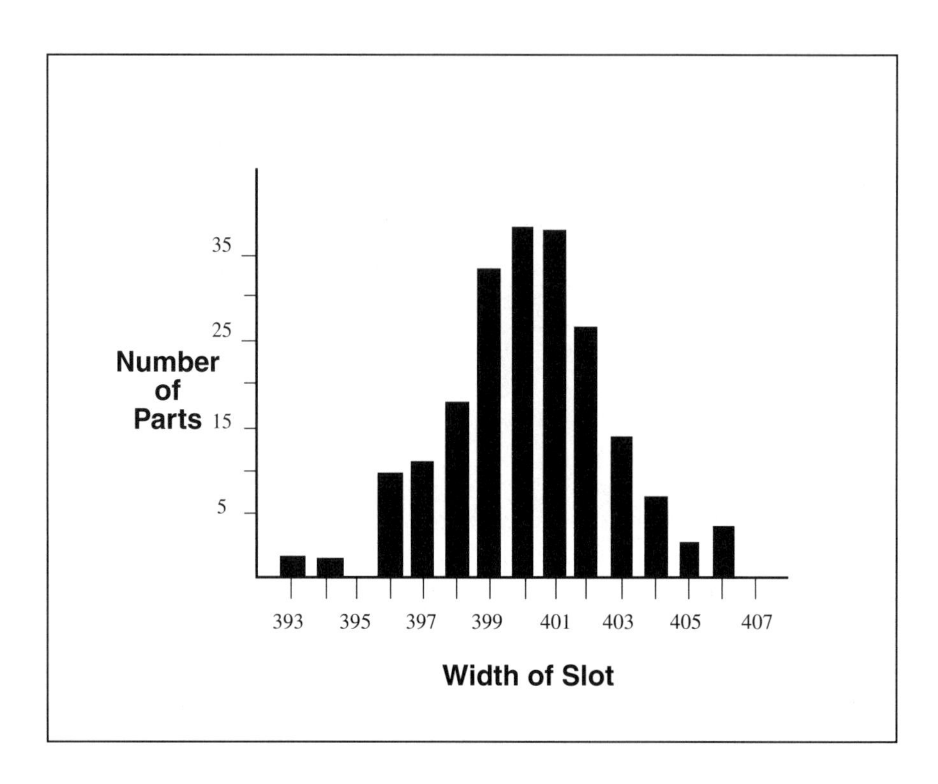

For NML, target = 400, mean = 398.62. So k = 1.38 and $C_{pk} = 1.54 \times (1 - 1.38/10) = 1.33$.

The Japanese team reviewed the results and explained to NML that, while a process capability index of $C_p = 1.5$ or $C_{pk} = 1.33$ had been acceptable in the 1980s (see Figure 10-7), a value of at least 2.0 was required now for this type of part. Therefore NML would have to improve the capability of the production process or the contract would not be renewed.

Notice that the variable in the numerator of C_p comes from product design, and the variable in the denominator comes from manufacturing. Consequently, C_p is a measure of the tightness of the fit between design and manufacturing. The higher C_p is, the easier it is for manufacturing to produce the design.

It is traditional to assume that the variation in a manufacturing process follows a normal distribution. The number of defects that would occur when $C_p = 1.0$ is 2700 ppm (parts per million). A $C_p = 2.0$ gives a level of quality of .002 ppm. The "Six Sigma Program" at Motorola and IBM is equivalent to $C_p = 2.0$. In this program, the variation in the process must be sufficiently small that the upper specification limit is at least six standard deviations (6σ) above the process mean, and the lower specification limit is at least 6σ below the process mean; that is:

$$C_p = \frac{\text{specification width}}{\text{6 standard deviations}} = \frac{6\sigma + 6\sigma}{6\sigma} = 2.$$

Another important reason for pursuing high process capabilities is that, in some manufacturing processes, such as those that run unattended, it is not unusual for the process mean to drift a little, as much as $\pm 1.5\sigma$. This drift will be noticed by the process control charts after a short time lag. Until it is noticed, however, the process may produce defective parts. The effect of a drift of 1.5σ on a process with $C_p = 1.0$ is a dramatic drop in quality. If $k = 1.5\sigma$, $C_{pk} = 1.0(1 - 1.5\sigma/3\sigma) = .5$, the number of defects jumps to 66,810 ppm. However, when $C_p = 2.0$, the same drift gives $C_{pk} = 2.0(1 - 1.5\sigma/6\sigma) = 1.5$, and the number of defects is 3.4 ppm. A high C_p guarantees a high level of quality even when the process experiences temporary problems.

Concurrent Engineering

The 40/30/30 rule states that about 40 percent of all quality problems, as perceived by the customer, result from poor product design, 30 percent result from errors made during manufacturing,

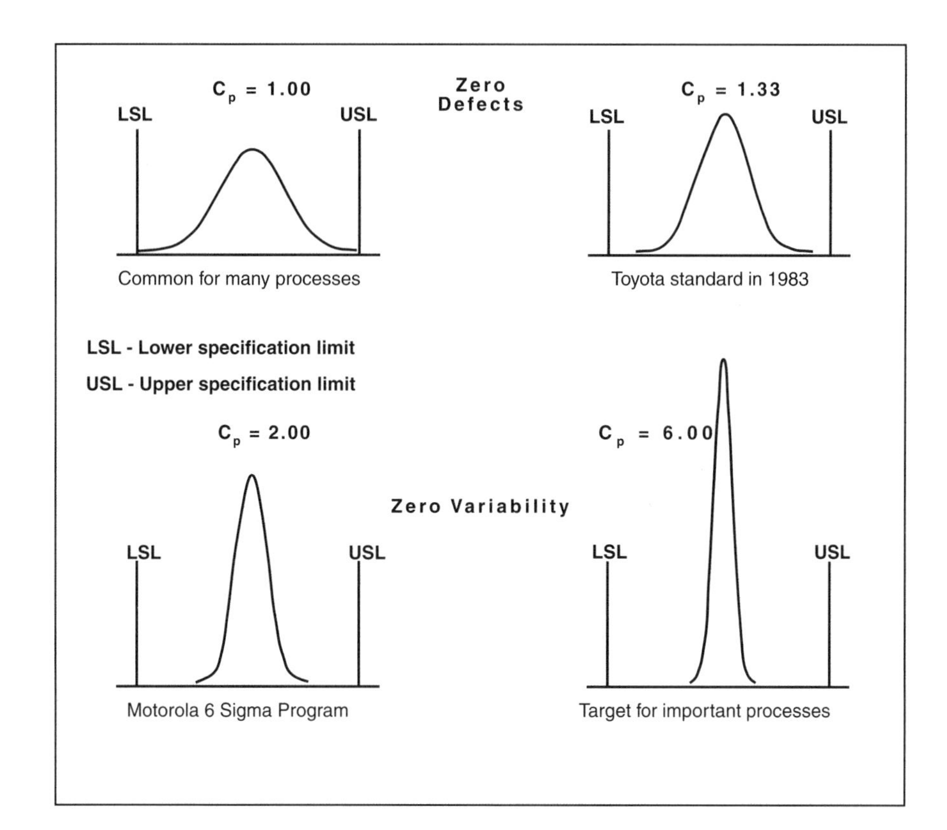

FIGURE 10–7

Values of the Process Capability Index

and 30 percent result from defective materials purchased from suppliers. Concurrent engineering is a soft technology that focuses on problems in the first category. It recognizes that improvements are easiest to make when the design of the product is still on paper. Concurrent engineering teams are multifunctional, with members from marketing, product design, production, finance, and so on. The goals of the team are to 1) design a product that is easy to manufacture, 2) use the best manufacturing processes to produce the product, and 3) reduce the time required to introduce a new product.

SITUATION 10.6

Concurrent Engineering at Motorola[4]

MOTOROLA once defined the product development cycle time as the time from conception of the product until the product was released to manufacturing. Today, the definition is the time from conception until the product is manufactured at the quality level required to satisfy the customer.

To improve the product development cycle, Motorola created a contract book and a five-step process for product development. The contract book sets specific, unchangeable deliverables for documentation, completion times, quality, reliability, and cost, and defines the relationship among product design, manufacturing, and marketing. The steps in the product development process are:

1. Determine the physical and functional characteristics of the product necessary to satisfy the customer.
2. Identify the key characteristics of the design that control the requirements determined in step 1.
3. Identify the process controlling each key characteristic.
4. Determine the capability index of the process used for each key characteristic.
5. If the capability index (C_p) is less than 2.0, seek design or process alternatives.

SITUATION 10.7

Northrop Implements Concurrent Engineering

MANAGEMENT at Northrop was concerned about its chances of landing a multibillion dollar contract for a new 1990s fighter plane. Northrop competed by providing order winning levels of performance and innovativeness, and market qualifying levels of cost, quality, and delivery. It was becoming difficult, however, to provide the required levels of cost and delivery because of the numerous engineering change orders to be handled. Northrop decided to implement a concurrent engineering program so that it could reduce the number of change orders. Northrop called its program "parallel release." Desks and people were moved so that designers and process engineers could work side by side. Designers accounted for process capabilities in their designs. Process engineers understood designers' problems and worked to make needed process improvements. The benefits came quickly. Change orders were reduced by 70 percent, and cost and delivery improved.

HARD TECHNOLOGIES

The development of new materials and the availability of cheap computing power produced the new generation of manufacturing equipment and processes that we call hard technologies (see Figure 10-8). Hard technologies are very expensive. They are relatively easy to implement—engineering purchases them and has them installed on the factory floor, the vendor helps with the startup, and they are handed over to the production department (or some other user). It is fair to say that many organizations have been disappointed with what these hard technologies finally provide in terms of market qualifying and order winning outputs.

Hard technologies must be appropriate for the manufacturing strategy and for the production system. They should be selected because they will help to provide high levels of the market qualifying and order winning outputs. They should not be implemented until manufacturing has been focused and appropriate soft technologies have been implemented.

FIGURE 10–8

Some Hard Technologies

NC (numerical control) equipment

Equipment having a control system that permits a program that, when read and executed, produces a part.

CNC (computer numerical control) equipment

NC equipment controlled by its own computer, which permits faster development, handling, and execution of NC programs, and provides more control over the equipment.

Machining center

An CNC machine capable of performing more than one type of machining operation. (A machine with the functionality of several different CNC machines.)

DNC (direct numerical control)

Multiple CNC equipment connected to the same supervisory computer, which permits better control of several pieces of equipment.

Robot

Programmable manipulator used to grasp and move parts while performing activities such as painting, welding, loading and unloading, and assembling.

AGV (automated guided vehicle)

Programmable device for moving materials around a factory.

CAD (computer aided design)

Design of a part and production of a part drawing on a computer.

CAE (computer aided engineering)

Design and engineering of a part, and production of the part drawing on a computer.

CAD/CAM (computer aided design / computer aided manufacturing)

Production of an NC program to manufacture a part, which was developed by CAD, on NC and CNC equipment.

MRP (manufacturing resources planning)

Computer and software used to generate production plans (for material, labor, and equipment), and monitor and control execution of the plans.

CIM (computer integrated manufacturing)

Using the computer to integrate all manufacturing activities, including order entry, product design, production planning, performing manufacturing operations, inspection, shipping, and billing.

SITUATION 10.8

Zepf Succeeds with Hard Technology

ZEPF TECHNOLOGIES is a world leader in the design and manufacture of feedscrews. Feedscrews are used to orient and shunt high volumes of fast-moving products in continuous flow production systems. For example, a single row of bottles moving at a high speed may enter a parallel set of rotating feedscrews, each about 20 cm in diameter and 150 cm in length, and emerge as an evenly spaced, double row of bottles ready to go into shipping containers. The geometry of today's feedscrews could not be designed or produced prior to the mid-1980s because the hard technologies were not available. At that time, Zepf Technologies purchased and developed the hard technology it needed to be an industry leader.

In 1980, when it had just surpassed the $1 million gross sales mark, Zepf spent $750,000 on CNC machines. No internal justification memos were necessary. To remain in business—to provide the market qualifying outputs—Zepf had to purchase this equipment. The CNC machines worked quickly. There were long idle periods, however, when the machines waited for the programs to run them because it took a long time to design the feedscrews, calculate the cutting path, and prepare the NC programs. In 1981, Zepf started to automate the design process. The goal was to present the designer with a series of prompts for the screw specifications. The program would take the specifications, calculate the cutting path, and prepare the program, all without the need for any intermediate drawings.

New programs were added as volumes increased. In 1986, when gross sales were $2.7 million, a new $200,000 computer was purchased. Soon programming and design activities occupied the time of 20 percent of the staff. By 1992, Zepf was on its fourth generation of computer aided engineering (CAE) software. The company developed its own CAE software and built its own CNC machines. Engineering, design, quality control, process planning, DNC, and CNC were fully integrated. Research and development spending was 12.5

percent of sales. As the following chart shows, remarkable improvements in cost, quality, performance, delivery, flexibility, and innovativeness were achieved.

Feedscrew Type	Year	Hours to Program	Hours to Machine
Simple	1981	32	5
	1991	0.5	0.5
Complex	1981	100	10
	1991	1	2

SUMMARY

Regardless of the improvement approach used, improvements should be made in a specific order. First, manufacturing is focused. Soft technologies are implemented next. Only then are appropriate (and by this time, obvious) hard technologies added. Unsuccessful attempts to improve manufacturing capability fail because one of these steps is missed or the steps are done in the wrong order.

When considering whether to implement a particular soft or hard technology, one must determine the manufacturing levers that will be affected by the new technology, decide if the changes to the levers are appropriate for the production system, and assess whether the new technology will improve the ability of the production system to provide the market qualifying and order winning outputs at the target levels.

NOTES

1. B. Wickens, "Lessons in How to Survive," *Mclean's Magazine*, Vol. 104, No. 32, p. 32. August 12, 1991.

2. Adapted from "TRW," *Target*, Vol. 1, No. 8, pp. 6–9, 1985.

3. This is adapted from a problem on page 754 in R. Schmenner, *Production and Operations Management*, 4th edition, New York: Macmillan, 1987.

4. Smith, B., "Six-Sigma Design," *IEEE Spectrum*, pp. 44–45, September 1993.

Further Reading

Hayes, R., and S. Wheelwright, Chapter 4, "Facilities Strategy," in *Restoring Our Competitive Edge: Competing Through Manufacturing*, New York: John Wiley & Sons, 1984.

Skinner, W., "The Focused Factory," *Harvard Business Review*, pp. 113–121, May–June 1974.

Venkatesan, R., "Cummins Engine Flexes Its Factory," *Harvard Business Review*, pp. 120–127, March–April 1990.

CHAPTER 11

LEARNING AND THE PRODUCT LIFE CYCLE

A product doesn't last forever. It is developed and introduced to the market. If it is successful, demand for it grows rapidly. Eventually, growth stops and it begins a stable, mature phase. This phase ends when customer preferences change or new products appear that render the product obsolete. These phases or stages constitute the *product life cycle*. The demands on manufacturing change as the product moves through these stages. Because the product design, production volume, and manufacturing outputs demanded by customers change, different production systems are required at different stages of the product life cycle. More and more units are produced as the product moves through its life cycle. This gives the organization manufacturing the product an opportunity to learn, which enables it to improve the product and the production system, resulting in higher levels of the manufacturing outputs.

Before taking up the effects of learning and the product life cycle on manufacturing strategy, consider the following situation where learning and the product life cycle were critical factors in one organization's takeover of the world market for a consumer product.

Situation 11.1

Samsung's Learning Curve in Producing Microwave Ovens*

In 1976, after one year's work, Samsung's first prototype microwave oven was finished and ready for testing. The chief engineer pushed the ON button. In front of his eyes, large plastic sections of the oven started to melt. More weeks of hard work followed before a second prototype was ready. This time, when the oven was turned on, another part melted. It was a discouraging moment. The Japanese and Americans were selling over 4 million microwave ovens a year, and Samsung Co. of Korea could not get even a single prototype to work.

In June 1978, another prototype was ready. This time nothing melted. The product was still too crude to compete in the world market, but at least it worked. There were hardly any orders for the new microwave oven, and so Samsung organized a job shop production system to build between one and five ovens each day. Production costs were high because of the low volume. Development cost was also high. Samsung's strategy for winning orders was to tailor its product to foreign tastes—to make unique models for unique markets. Samsung would provide high levels of flexibility and innovativeness, something that other manufacturers were not doing.

A year later, after 1,460 ovens had been produced, Samsung started its first sales push. Soon it had an order from Panama for 240 ovens. Within weeks, the product was redesigned to meet the customer's needs, the ovens were manufactured, and the order was shipped. The order helped Samsung learn what its customers wanted and gave the company enough confidence to apply for the Underwriters' Laboratory (UL) approval—something it needed if it wanted to export to the United States. Late in 1979, Samsung received the UL approval.

In 1980, J.C. Penney, one of the largest retailers in the United States, asked Samsung if it could produce a

* Reprinted by permission of the *Harvard Business Review*. An excerpt from "Fast Heat: How Korea Won the Microwave War," by Ira Magaziner and Mark Patinkin, January/February 1989.

microwave oven that could be sold in the United States for $299. At the time, microwave ovens were selling for between $350 and $400. J.C. Penney had been looking for an oven it could sell for less, but hadn't found one with U.S. or Japanese manufacturers. Samsung accepted an order from J.C. Penney for a few thousand units. A team consisting of factory and product design employees was formed. At Samsung, production was most important, so the product had to be designed with production in mind. The new J.C. Penney microwave oven and the new production system were designed concurrently. A batch flow production system was organized to produce between 10 and 15 ovens per day. Ovens were produced by day, and during the night improvements were made to the production process. Improvements were also made at many of the vendors that were part of the project. Before long, production increased to 60 ovens a day, which was enough to meet the J.C. Penney order. J.C. Penney liked the ovens and soon asked for 200, then 250 ovens per day. When the volume increased and the design stabilized, the production system was changed to an operator-paced line flow production system.

By 1982, Samsung produced 200,000 microwave ovens a year. The big manufacturers in Japan and the United States, who produced over 5 million ovens annually, noticed and started to lower their prices. Samsung realized that if it was to keep growing, it had to lower its prices even more. Samsung's managers scrutinized their cost structure. They found two ways to lower costs: 1) increase vertical integration backward to the production of all important components, and 2) increase volume so that the production system could be changed to an equipment-paced line flow.

1. Increase Vertical Integration

The highest cost item in Samsung's oven was the magnetron tube, which generated the microwaves. It was the most complex device in the oven and Samsung did not have the expertise to produce it. The company bought the magnetron tube from a Japanese supplier. Costs would be significantly lower if Samsung produced its own magnetron tubes. The company first approached producers in Japan for technical assistance but was turned down. That left Amperex—the only U.S. manufacturer of magnetron tubes. However, Amperex was leaving the magnetron tube business because it could not compete with Japanese producers. Samsung quickly bought Amperex's production process. In late 1982 and early 1983, the Amperex equipment for producing magnetron tubes arrived in Korea. Before long, Samsung was manufacturing its own magnetron tubes.

2. Implement an Equipment-Paced Line Flow Production System

If Samsung could obtain more orders, volume would increase and the production system could be changed to an equipment-paced line flow system, which would lower production costs. In 1983, General Electric (GE) decided to begin sourcing small and midsize microwave ovens from Asia and manufacture only full-size models in the United States. The biggest orders went to Japan, but a small order for 15,000 ovens went to Samsung. At first, Samsung's ovens did not meet GE standards. But with the help of GE's quality engineers, they improved. GE was impressed and more orders followed. To handle the higher volume, Samsung's production system was changed to an equipment-paced line flow production system.

Samsung continued to learn and improve its manufacturing capability. In 1984, it had four equipment-paced assembly lines. Specially designed microwave ovens were soon exported to Germany, France, and Scandinavia. In May 1985, General Electric announced it would stop U.S. production of microwave ovens. From now on, GE would do sales and service, and Samsung would do all the manufacturing. GE realized that its manufacturing capability was so far behind competitors like Samsung that it could not compete (see Figure 11-1).

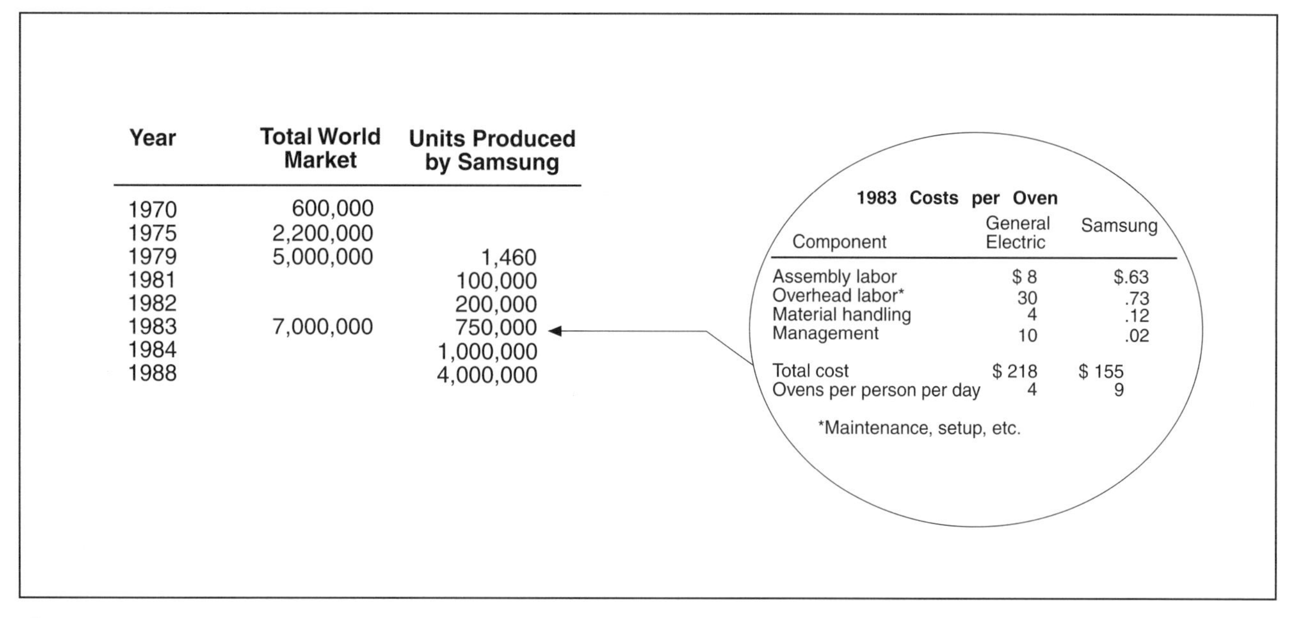

Year	Total World Market	Units Produced by Samsung
1970	600,000	
1975	2,200,000	
1979	5,000,000	1,460
1981		100,000
1982		200,000
1983	7,000,000	750,000
1984		1,000,000
1988		4,000,000

1983 Costs per Oven

Component	General Electric	Samsung
Assembly labor	$ 8	$.63
Overhead labor*	30	.73
Material handling	4	.12
Management	10	.02
Total cost	$ 218	$ 155
Ovens per person per day	4	9

*Maintenance, setup, etc.

Source: Magaziner and Patinkin 1989.

FIGURE 11–1

Production Volume and Cost at Samsung and General Electric

LEARNING

Learning was a critical factor in Samsung's success. Each time Samsung learned, the level of manufacturing capability increased and higher levels of the manufacturing outputs—cost, quality, performance, delivery, flexibility, and innovativeness—were provided. Learning occurred continuously. Samsung learned about the product and the production process when it developed prototypes. The company learned from its customers in Panama, at J.C. Penney, and at General Electric; from its suppliers; and from equipment vendors like Amperex. Learning resulted in better products and better processes.

Learning always coincided with increases in production volume; it either resulted in an increase in volume or occurred as a result of an increase in volume. Learning was not left to chance. Samsung developed formal systems to ensure that learning occurred. Teams consisting of product design and factory employees were organized to develop new models and were sent to other countries to learn from customers and manufacturers. Engineers were encouraged to try new ideas.

The effect of learning is shown on a graph called the learning curve. The cost of a product is plotted against the total production volume of the product up to that point in time. (Sometimes price is used as a proxy for cost.) The graph usually shows a product cost that decreases as the production volume increases. The rate of decrease is called the slope of the learning curve. Part 1 of Figure 11-2 shows the learning curve for the integrated circuit. In 1964, after 2 million integrated circuits had been produced, the price was about $12 per unit. By 1972, almost 2 billion units had been produced and the price had dropped to about 70 cents per unit. When both the price per unit and the total production are plotted on logarithmic scales, the learning curve appears to be linear. The slope of this learning curve is 75 percent. The slope, L, is interpreted as follows. Each time the total cumulative production doubles the price drops by $100 - L$ percent. In the case of the integrated circuit, $L = 75$, so the price of an integrated circuit drops by $100 - 75 = 25$ percent each time the cumulative production doubles. Note that cost does not drop automatically as production volume increases. Cost drops only if improvements are made. Approaches like TQM, SCM and kaizen generate the improvements (see Chapter 9).

Part 2 of Figure 11-2 shows how the price of the Model T Ford fell when Henry Ford made his legendary improvements at the River Rouge plant (see Situation 9.6 on page 175). The slope of this learning curve is 85 percent. Another example of the effect of learning is shown in Part 3 of the figure. The slope of the learning curve for limestone was 80 percent over the period from 1929 to 1971. Notice that the price was $1.70 per ton in the mid-1940s when the cumulative production was 3 billion tons. When the total production doubled to 6 billion tons, the price dropped to $1.70 × 80 percent, or $1.36. Therefore, if management continues to make improvements, a target for the price when the cumulative production doubles again to 12 billion tons is $1.36 × 80 percent, or $1.09 per ton.

The most dramatic examples of learning have occurred recently in the electronics industry. The prices of calculators, electronic watches, televisions, and computers have dropped dramatically in the years following their introduction. Part 4 of Figure 11-2 presents some data for the RAM chip. The price is in thousandths of a cent per bit of memory. In 1978, the 1 trillionth chip was manufactured and the price was .075 cents per bit. Within three years, 10 trillion chips had been manufactured and the price had dropped to .015 cents per bit. The slope of the learning curve was a remarkable 62 percent.

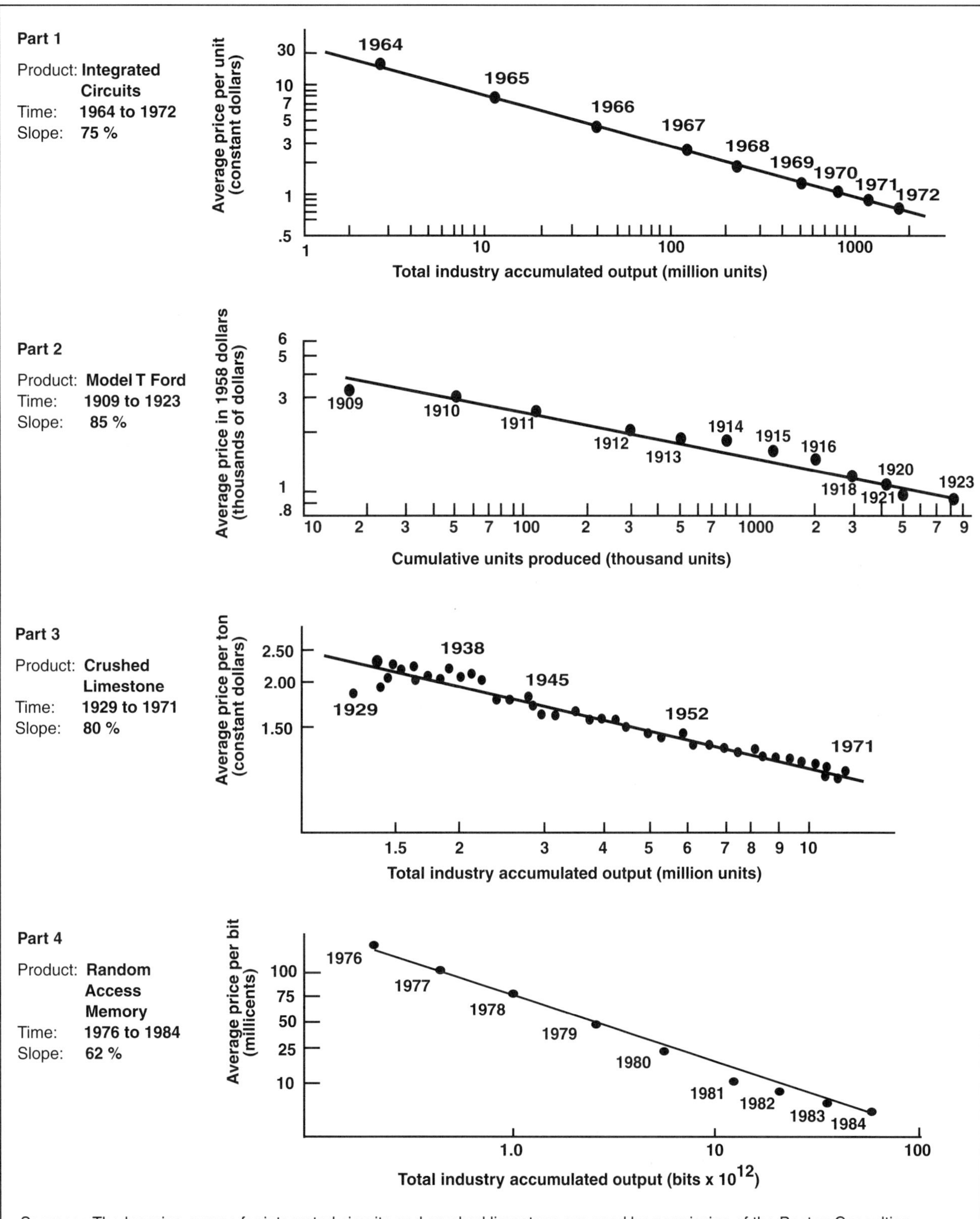

Sources: The learning curves for integrated circuits and crushed limestone are used by permission of the Boston Consulting Group, Inc., Boston Masachusetts. The learning curves for the Model T Ford and random access memory are reprinted by permission of the *Harvard Business Review* from exhibits in "Limits of the Learning Curve" by William Abernathy and Kenneth Wayne, September/October 1974, and "Building Strategy on the Experience Curve" by Pankaj Ghemawat, March/April 1985 (copyright © 1974 and 1985 by the President and Fellows of Harvard College; all rights reserved).

FIGURE 11–2
Four Well-Known Learning Curves

FIGURE 11–3

Using Learning Curves to Set Targets for Improvements

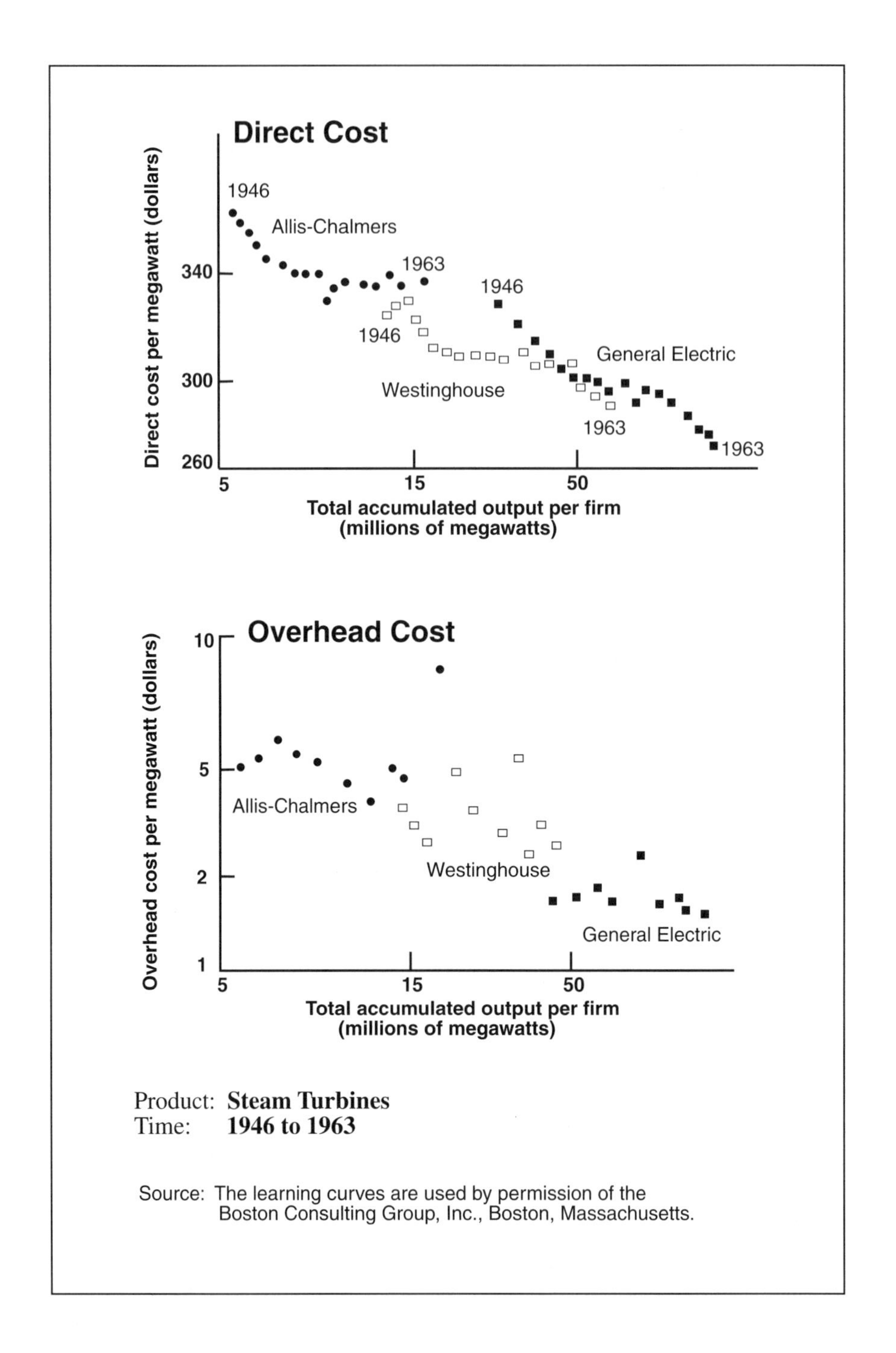

Product: **Steam Turbines**
Time: **1946 to 1963**

Source: The learning curves are used by permission of the Boston Consulting Group, Inc., Boston, Massachusetts.

Cost data for three companies manufacturing steam turbines are shown in Figure 11-3. The first graph shows the effects of learning on the direct cost of manufacturing this product. The second graph shows the effects of learning on the overhead cost. Both direct cost and overhead cost decreased as the number of units produced increased (measured as the total cumulative megawatts of power of the steam turbines) for all three companies. The slopes of the learning curves are similar for the three companies, which suggests that each was making similar improvements. The slopes for overhead cost are not as steep as for direct cost, suggesting that not as much learning is occurring in the activities that constitute overhead cost. In such a situation, management might use an improvement approach like SCM to see if it can increase learning in these activities.

Learning Formula

Let the slope of the learning curve be L percent, $0 \leq L \leq 100$. The cost per unit decreases by $100 - L$ percent when the number of units manufactured doubles. Mathematically, this means:

$$C_n = C_1(n^b)$$

where $b = \log(L/100)/\log 2$

C_1 = cost to manufacture the first unit

C_n = cost to manufacture the nth unit

Situation 11.2
Using the Learning Curve to Set Improvement Targets

Example 1

Problem

The price of a RAM chip (Part 4 of Figure 11-2) in 1978, when the 1 trillionth bit was manufactured, was 0.075 cents per bit. By 1984, some 80 trillion bits had been manufactured. What should the price be in 1984 if the slope of the learning curve is 62 percent? 50 percent?

Solution

Let C_1 be the price for the one trillionth bit: $C_1 = .075$

C_{80} be the price for the 80 trillionth bit

L = 62 percent:

$$b = \log(62/100)/\log 2 = -.69$$

$$C_{80} = C_1(n^b) = .075(80^{-.69}) = .004$$

$L = 50$ percent:

$$b = \log(50/100)/\log 2 = -1.0$$
$$C_{80} = .075(80^{-1.0}) = .0009$$

The price per bit is more than four times higher when the slope of the learning curve is 62 percent than when it is 50 percent.

Example 2

Problem

The learning curve for integrated circuits was shown in Part 1 of Figure 11-2. Suppose that in 1972 a new competitor wanted to enter the industry. The competitor estimated that it would take two years to develop the appropriate products and production system. What target should the competitor set for its cost to manufacture an integrated circuit in 1974?

Solution

The cumulative industry volume in 1974 can be forecast from the actual volumes in 1964 through 1972. An estimate of 3.2 billion units in 1974 was obtained from a simple forecasting model.

Let $C_{1.8}$ be the actual price when the 1.8 billionth unit was manufactured (sometime in 1972): $C_{1.8} =$ \$0.70

$C_{3.2}$ be an estimate of the price when the 3.2 billionth unit is manufactured in 1974.

The slope of the learning curve is $L = 75$ percent. Therefore:

$$b = \log(75/100)/\log 2 = -.415$$
$$C_{1.8} = 0.7 = C_1(1.8^{-.415}) \text{ or } C_1 = .89$$
$$C_{3.2} = C_1(3.2^{-.415}) = .89(3.2^{-.415}) = .55.$$

The new competitor's target cost in 1974 should be less than the anticipated price of \$0.55.

It is traditional to link learning to a particular product and the total production volume of that product from the time the product was first introduced. Since the production volume changes as a product progresses through its life cycle, the amount of learning that occurs varies during each stage in the product life cycle.

Product Life Cycle

Product sales follow a pattern over time (see Figure 11-4). The pattern is called the product life cycle, which consists of six stages.

Development Stage

A new product is designed, prototypes are developed and tested, and small volumes of the new product are produced by skilled workers on general-purpose equipment in a job shop production system. Design costs and production costs are high. Marketing tries to gain acceptance for the product from distributors and customers.

Growth Stage

Orders are obtained. The product design is changed to satisfy the requirements for each order. Changes are made to the production process so that the new designs can be manufactured. As more orders are obtained, manufacturing struggles to produce increasing volumes of customized parts. The job shop production system is changed to a batch flow system.

Shakeout Stage

The market becomes more competitive, and standard designs and features appear. Customer expectations for quality, performance, and flexibility rise. Companies that cannot provide market qualifying levels of the manufacturing outputs drop out. Production systems improve so that the required levels of the manufacturing outputs can be provided. Some competitors find market niches and begin to specialize.

Maturity Stage

Standard industrywide designs emerge. Customers require very high levels of the cost and quality manufacturing outputs. Production volume is high and grows at about the same rate as the economy. Production is moved to a line flow production system.

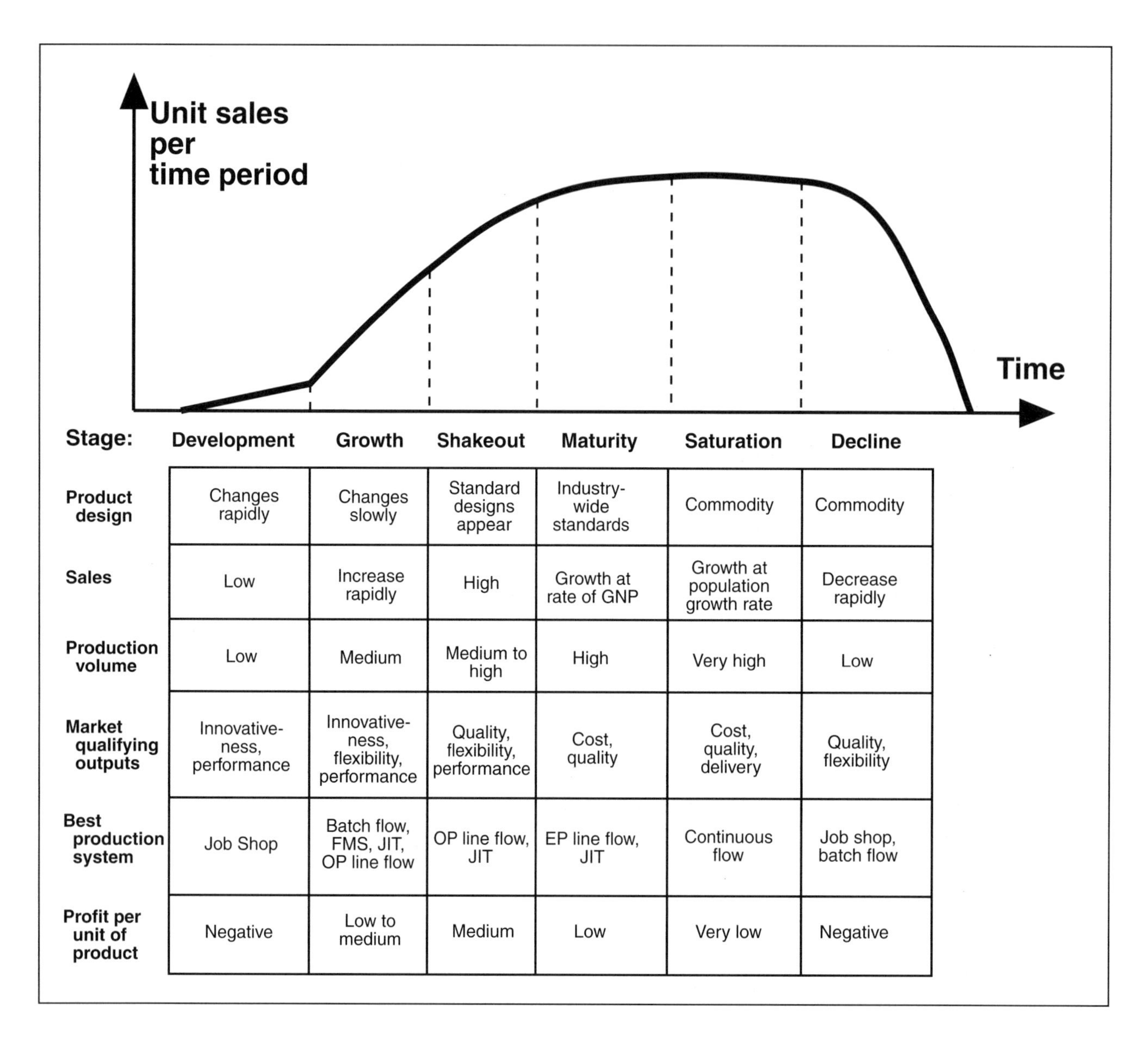

Stage:	Development	Growth	Shakeout	Maturity	Saturation	Decline
Product design	Changes rapidly	Changes slowly	Standard designs appear	Industry-wide standards	Commodity	Commodity
Sales	Low	Increase rapidly	High	Growth at rate of GNP	Growth at population growth rate	Decrease rapidly
Production volume	Low	Medium	Medium to high	High	Very high	Low
Market qualifying outputs	Innovative-ness, performance	Innovative-ness, flexibility, performance	Quality, flexibility, performance	Cost, quality	Cost, quality, delivery	Quality, flexibility
Best production system	Job Shop	Batch flow, FMS, JIT, OP line flow	OP line flow, JIT	EP line flow, JIT	Continuous flow	Job shop, batch flow
Profit per unit of product	Negative	Low to medium	Medium	Low	Very low	Negative

FIGURE 11–4

Stages in the Product Life Cycles

Saturation Stage

The product becomes a commodity and competes on the basis of cost, quality, and delivery. Production volume is high and grows slowly. Continuous flow production is used.

Decline Stage

New products appear and orders for the product drop rapidly. Customer preferences change. Soon the volume is so low that the dedicated production system cannot be used adequately. Production is moved to a batch flow or job shop production system.

Examples of products at different stages in their product life cycle are given in Figure 11-5. The length of, and the sales at, each stage, as well as the overall length of the life cycle, vary from product to product, and depend on factors such as the rate of technological change, the amount of competition in the industry, and customer preferences. Product life cycles are generally shorter today than they were in the past. Newer, better products are designed and manufactured at a faster rate than ever before; that is, the levels of performance and innovativeness are higher than ever.

The best production system for a product changes as the product moves through its life cycle. The relationship between the product life cycle and the production systems is shown in Figure 11-6. The job shop and batch flow production systems are normally used during the development stage. The FMS production system can be used during the latter part of this stage. During the growth stage, the production system changes to the operator-paced line flow system. The JIT production system can be introduced during the growth stage and used during the shakeout and maturity stages. An operator-paced line flow production system is normally used during the shakeout stage. As the product enters the maturity stage, the production system is changed to an equipment-paced line flow system. The continuous flow production system is best during the saturation stage, when the product is a commodity. An attractive feature of both the JIT and the FMS production systems is that they can be started much

FIGURE 11-5

Products at Different Stages in Their Product Life Cycle

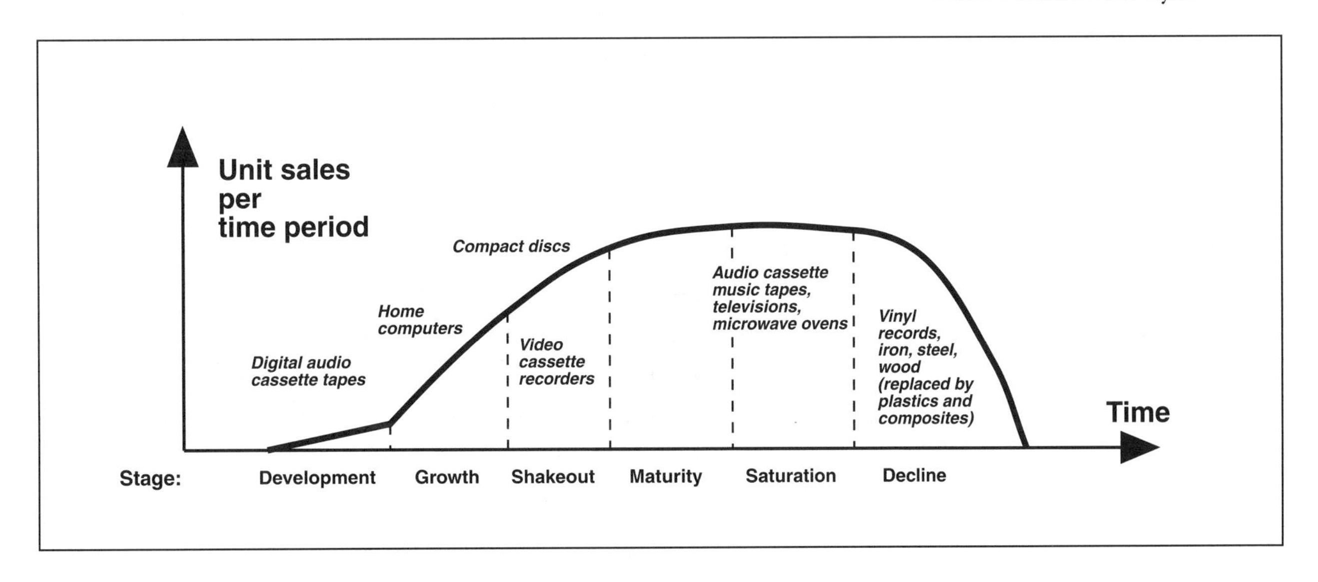

FIGURE 11–6

Product Life Cycle and the Production System

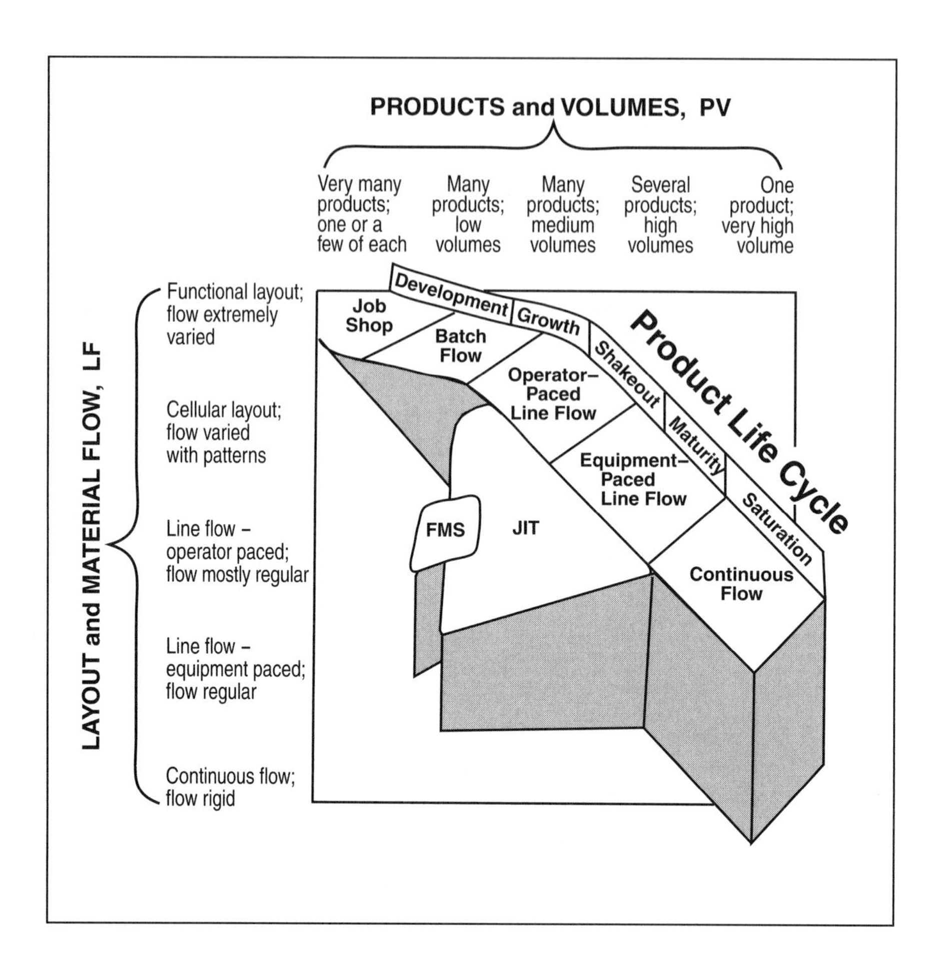

earlier in the product life cycle and used for a much longer period of time than any of the traditional line flow production systems.

The product life cycle for Samsung's microwave ovens is shown in Figure 11-7. Notice how often and how quickly Samsung changed its production system. Notice also that Samsung always used traditional production systems, never the difficult JIT and FMS systems.

SITUATION 11.3

Texas Instruments' and Hewlett-Packard's Comfort Zones on the Product Life Cycle[1]

TEXAS INSTRUMENTS (TI) and Hewlett-Packard (HP) are two companies whose manufacturing strategies in the early 1980s depended greatly on the product life cycle. TI was most comfortable when its market qualifying outputs were cost, quality, and delivery. As soon as TI developed a new product, it moved quickly toward line flow and continuous flow production systems,

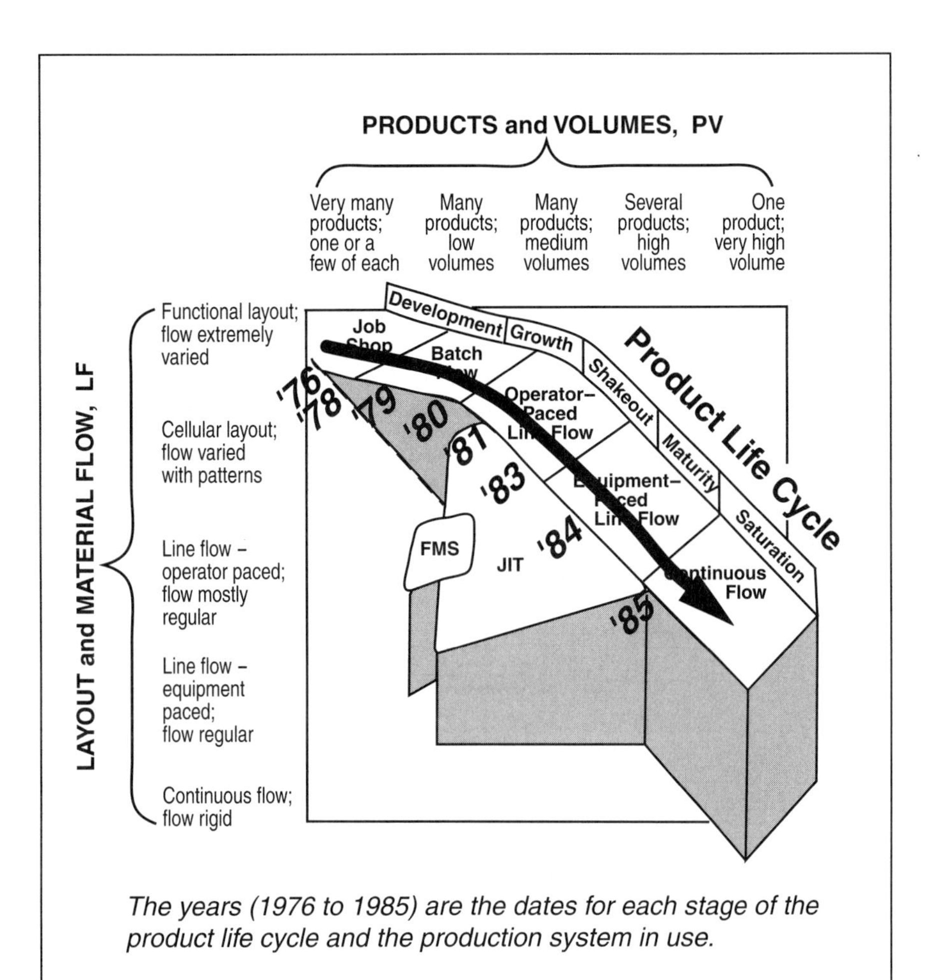

The years (1976 to 1985) are the dates for each stage of the product life cycle and the production system in use.

FIGURE 11–7

Product Life Cycle for Samsung's Microwave Oven

where it was easy to provide high levels of these outputs (see Figure 11-8). Cost was TI's order winning output, so the company strove to continually reduce the cost of its products. These reductions resulted in rapid increases in volume, which were necessary to adequately utilize TI's line flow and continuous flow production systems. Learning was very important (see Figure 11-9).

HP, on the other hand, was comfortable when its market qualifying outputs were performance, flexibility, and innovativeness. Consequently HP liked to develop new products and keep them until they reached the maturity stage of their product life cycles (see Figure 11-8). HP sought niche markets where customers wanted

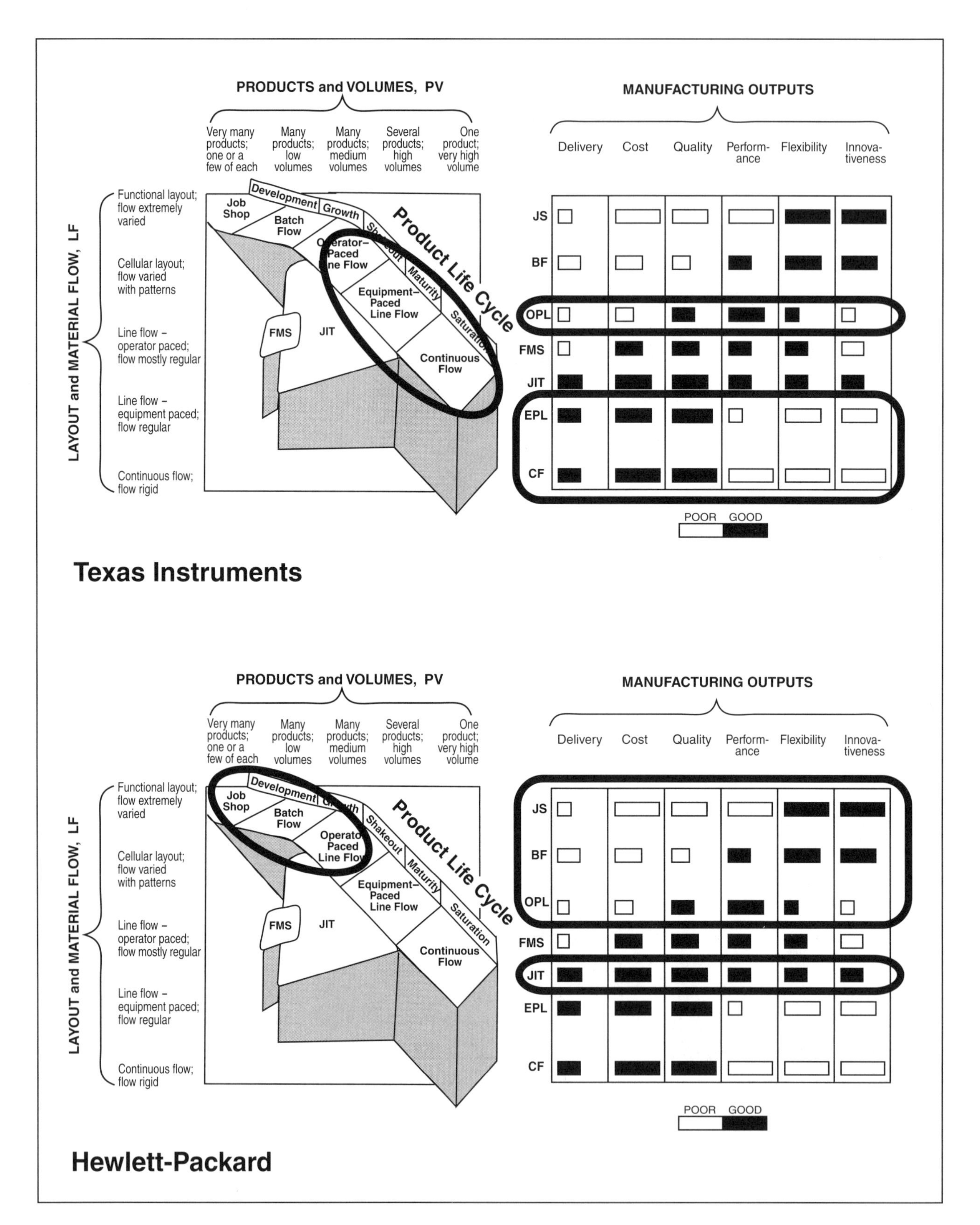

FIGURE 11–8
Product Life Cycle and Manufacturing Strategy at TI and HP

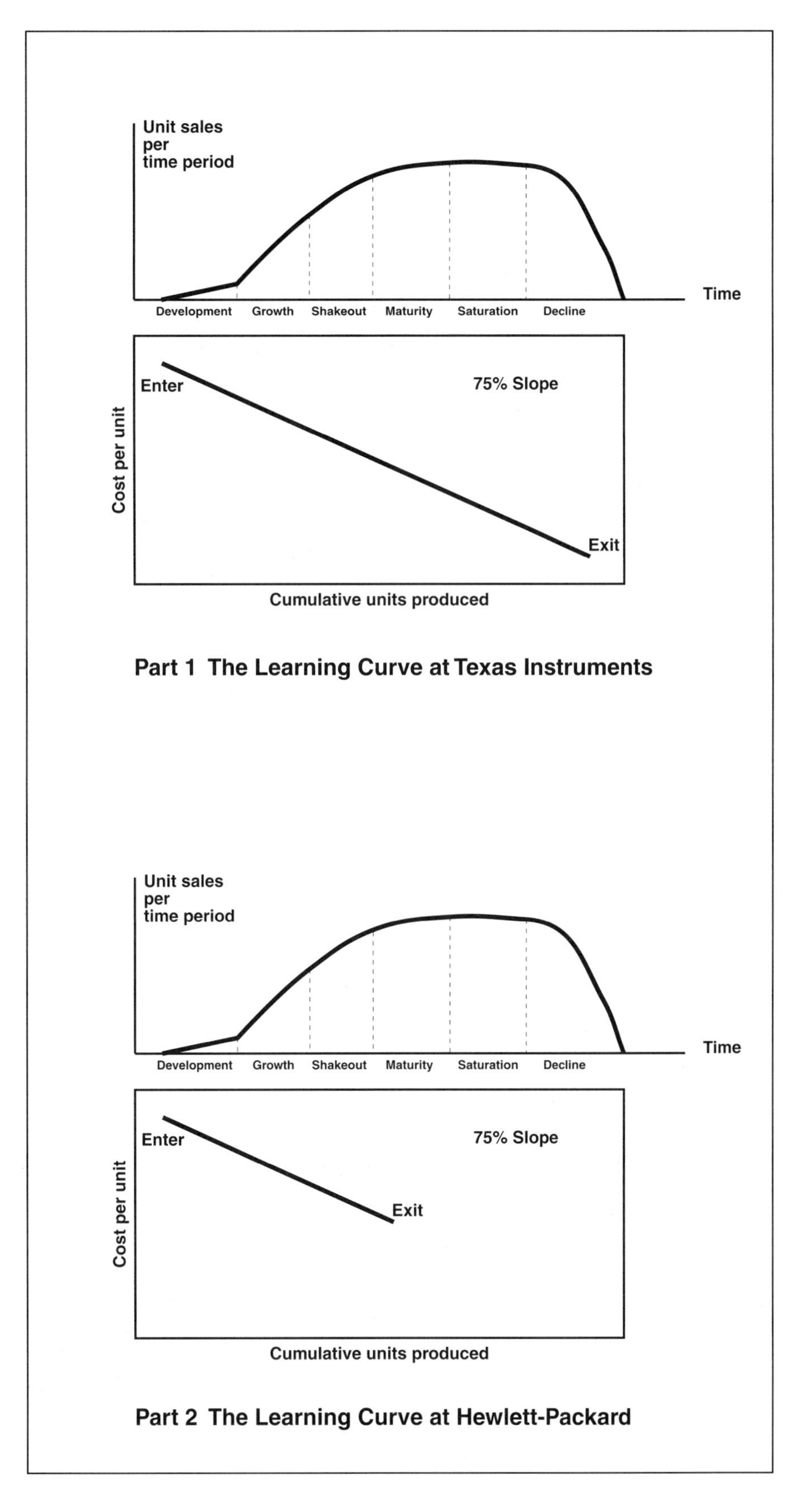

FIGURE 11–9

Learning at TI and HP

specialized products. HP exited these markets when the products matured, and competitors entered the market and changed the basis of competition to cost. Learning effects at HP were not as large as at TI because HP's volumes were never as high as TI's (see Figure 11-9).

Note that HP's manufacturing managers were North American pioneers with JIT production systems. There were two reasons for this: 1) HP produced many products in low to medium volumes, and JIT was an efficient production system for this mix; and 2) whereas corporate strategy emphasized performance, flexibility, and innovativeness, manufacturing managers in HP plants were evaluated on the basis of cost, quality, and delivery. The JIT production system was the only production system that could provide high levels of all these outputs. TI, on the other hand, was less interested in JIT production systems. TI used traditional line flow and continuous flow production systems, which provided the highest possible levels of the outputs that were important at TI—cost, quality, and delivery.

Summary

As a product moves through its life cycle, more and more units of the product are manufactured. This provides the organization with an opportunity to learn. Learning enables the organization to improve the product and the production system, which permits higher levels of the manufacturing outputs to be provided. The effects of learning are tracked on a learning curve. Learning is the theoretical underpinning for continuous improvement. Learning does not occur automatically. The manufacturer must seek improvements by, for example, using the improvement approaches discussed in Chapter 9. It is traditional to link learning to a particular product and the total production volume of that product from the time the product was first introduced. Because the production volume changes as a product progresses through the stages of its life cycle, the amount of learning that occurs varies during each stage. The best production system changes as the product moves through its life cycle.

NOTES

1. Adapted from S. Wheelwright, "Strategy, Management, and Strategic Planning Approaches," *Interfaces*, Vol. 14, No. 1, pp. 19–33, 1984.

FURTHER READING

Abernathy, W., and K. Wayne, "Limits of the Learning Curve," *Harvard Business Review*, pp. 109–119, September–October 1974.

Magaziner, I. and M. Patinkin, "Fast Heat: How Korea Won the Microwave War," *Harvard Business Review*, pp. 83–92, January/February 1989.

CHAPTER 12

EVALUATION OF INVESTMENTS IN MANUFACTURING

The capital appropriation process is often a source of frustration for manufacturing managers as well as financial managers. Manufacturing managers can't always convince financial managers that their proposals for investments in manufacturing are financially sound. Financial managers can't convince manufacturing managers that their rejections of some proposals are in the best long-term interests of the organization. The following incident is typical of the frustration that surrounds this process. According to a manufacturing manager at a Fortune 500 company:

> I can take $750,000 out of this region's inventories, and still maintain customer service levels, by spending $500,000 to retrofit a production process with modern manufacturing technologies. Common sense tells me that this is a sound investment—it costs the company nothing—but the financial people won't give me the money because they say the project has a four-year payback.

Who is right here? Before answering this question in Situation 12.5, we will review the criteria used to assess proposed investments in manufacturing. There are six interrelated criteria, three strategic and three economic. The criteria and the sequence in which they should be applied are shown in Figure 12-1.

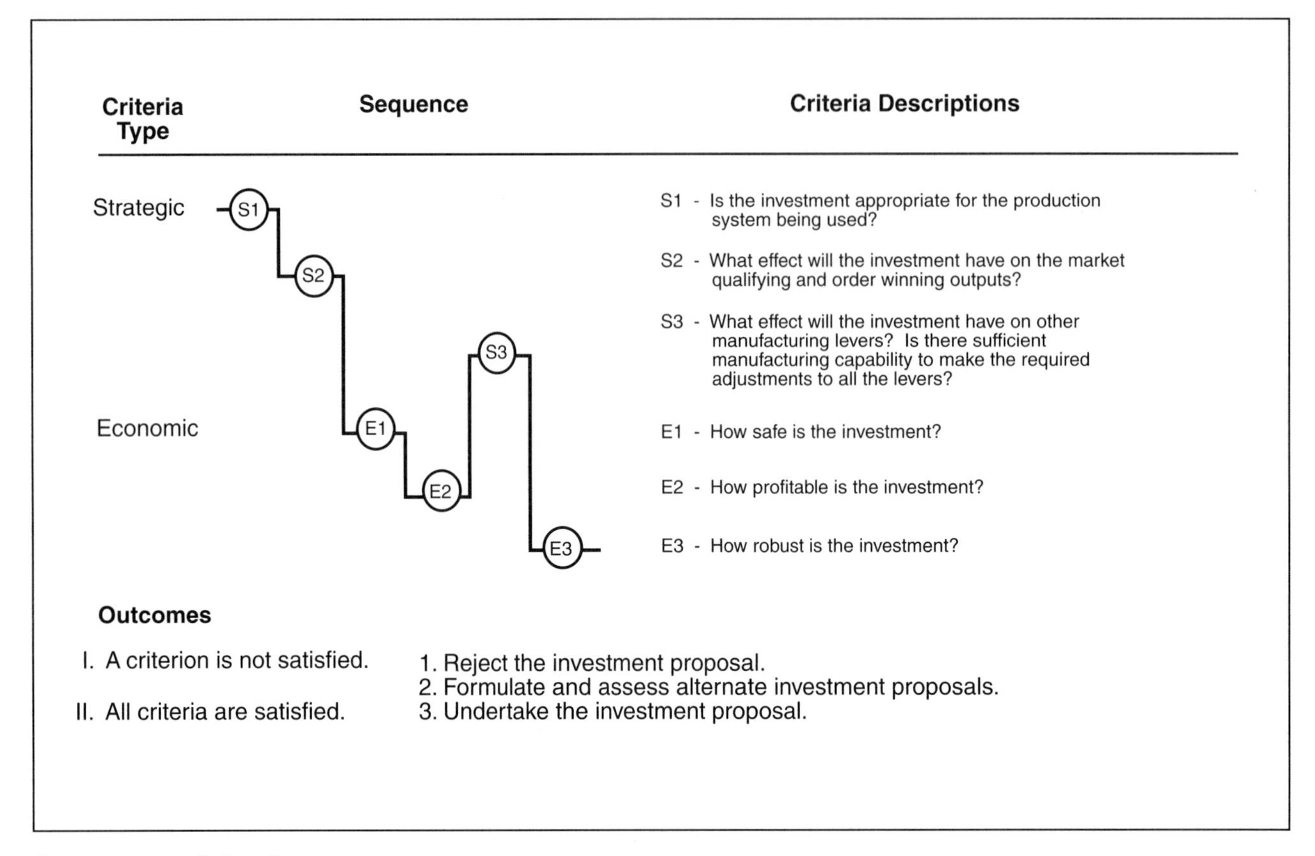

FIGURE 12–1

The Assessment Process for Evaluating Investments in Manufacturing

Applying Strategic and Economic Criteria

The strategic criteria indicate whether the proposed investment is consistent with the manufacturing strategy, and therefore with the corporate and business unit strategies (Chapter 8). Each of the three strategic criteria is a check of an element of the manufacturing strategy; that is, it checks the fit between the proposed investment and one of the elements on the manufacturing strategy worksheet. The economic criteria gauge the financial benefits of the proposed investment, in the full knowledge that most of these benefits will be realized in the future and that predicting future benefits is a tricky business. The assessment process stops when a proposed investment fails to satisfy a criterion, in which case the proposal is rejected. If all six criteria are satisfied, then the proposal is accepted. When a proposal is rejected, the insights developed during the assessment process often are used to develop new and better proposals.

STEP 1: STRATEGIC CRITERION S1

Is the Investment Appropriate for the Production System Used?

Manufacturers have access to many new soft and hard technologies. Only those appropriate for the production system used should be considered. The amount of new technology in the proposed investment is also considered in light of the organization's attitude toward new technology. This attitude is related to the organization's level of manufacturing capability. Organizations with capability at or above the adult level tend to be technology leaders. Organizations with capability near the industry average level tend to be technology followers—that is, those who adopt new technologies only after others have adopted them and the extent of the benefits is well known.

STEP 2: STRATEGIC CRITERION S2

What Effect Will the Investment Have on the Market Qualifying and Order Winning Outputs?

The proposed investment is evaluated for its ability to narrow the gap between the current and the target levels of the market qualifying and order winning outputs. Will the proposed investment enable manufacturing to meet its targets for cost, quality, performance, delivery, flexibility, and innovativeness, or does it provide outputs that are not needed?

SITUATION 12.1

Welding Machine Producer Considers Two Improvement Options

A WELDING machine manufacturer used an operator-paced line flow production system to provide a market qualifying level of cost and an order winning level of quality. An equipment-paced line flow production system would provide better levels of cost and quality but could not be used because the product mix and volume—many different models in medium volumes—was not appropriate.

The manufacturer was considering two investment proposals, only one of which could be accepted. Its engineering group requested funds to upgrade an old CAD system to a modern CAD/CAM system that would reduce the time required to prepare drawings and CNC tapes. It would also help the engineering group standardize its designs and thereby reduce the number of part numbers. The investment would shorten delivery times and reduce cost. The production group, on the other hand, wanted to replace an old punch press with a modern CNC machine complete with automated material handling, which would reduce cost and increase quality.

The company decided that the second proposal satisfied strategic criterion S2 and the first proposal did not. Investing in the CNC punch press would help the company provide the market qualifying and order winning levels of the cost and quality outputs more than would the CAD/CAM system.

Situation 12.2

Plastic Packaging Manufacturer Accidentally Changes Production Systems[1]

A MANUFACTURER of plastic packaging materials wanted to upgrade the equipment used in its operator-paced line flow production system. The production system provided an order winning level of performance for the many products it produced. The investment in new equipment would cost $6 million and had a payback of 5.5 years. However, the corporation required a payback of less than four years for this type of investment. To reduce the payback to four years, production volume would have to increase by 50 percent. Manufacturing agreed to attempt this, and the $6 million proposal was approved.

Changes were made to win more orders. Some products were standardized and their prices were dropped. This produced more orders, but not without creating problems. Delivery time started to slip because of problems in scheduling the standard, high-volume products with the regular, nonstandard, medium volume products. Profit margins started to shrink because the low prices on the standard products were barely able to cover their cost of production. Other changes were planned to correct these problems.

Unfortunately the effect of these changes was to move the production system from an operator-paced to an equipment-paced line flow system and alter the order winning output from performance to cost, without ever making an explicit decision to change the manufacturing strategy. The investment proposal should not have been approved because it did not satisfy strategic criterion S1 or S2.

STEP 3: ECONOMIC CRITERION E1

How Safe Is the Money?

This step is the first check of an economic nature. It gauges the effect of uncertainties external to the proposed investment. For example, problems may develop that have the effect of turning what is currently an attractive investment into an unattractive one. New opportunities may appear that are much more attractive than the proposed investment. It is traditional to assume that the longer it takes for an investment in manufacturing to start generating a profit, the greater is the risk posed by these external uncertainties. The payback period is a measure of this risk. It is the time required for an investment to return earnings equal to the cost of the investment. The shorter the payback period, the sooner the organization will recover its money and the sooner it will be able to reinvest that money in new projects.

Many organizations require that the payback period for an investment be less than a specified maximum period, which depends on the type of investment, the scarcity of funds, and the rate of change of products and processes in the industry. For example, in the early 1980s most companies in the semiconductor industry required paybacks of less than two years on all investments in manufacturing. These companies were growing rapidly and investment funds were in short supply. In addition, the rate of product change and process change was very high in the industry.[2] In the late 1970s, many companies in the automotive industry required paybacks of five years or less on all investments in energy conservation projects. This lengthy payback period reflected the feeling at the time that the risk from external uncertainties was low for these projects.

Step 4: Economic Criterion E2

How Profitable Is the Investment?

The second economic criterion assesses the ability of the proposed investment to generate earnings. The larger the stream of earnings relative to the cost of the investment, the more attractive is the investment. When assessing the profitability of an investment, account is taken of the timing of the earnings. Obviously, the sooner earnings are received, the better. Three well-known measures of the profitability of an investment are return on investment, internal rate of return, and net present value.

Return on investment (ROI) is calculated as follows:

$$\text{ROI} = \frac{\text{Average annual net cash flow}}{\text{Initial investment}}$$

Internal rate of return (IRR) is the discount rate that equates the present value of the stream of cash flows with the initial investment. The net present value (NPV) is a current dollar amount equivalent to the stream of cash flows. NPV discounts the value of future cash flows to reflect the time value of money. The discount rate is the rate of return required for a project with a certain amount of risk. It is the return that the organization could expect to receive elsewhere for an investment of comparable risk. Obviously, the discount rate must exceed the organization's cost of funds.

Regardless of which measure of profitability is used, companies require that a proposed investment exceed a minimum value of the profitability measure. The minimum value depends on the profitability associated with alternative investments available to the organization.

SITUATION 12.3

Stamping Plant Assesses Investment in New Equipment

A STAMPING PLANT used a batch flow production system to produce metal stampings for the automotive industry. The plant wanted to replace an old press with a new 250-ton, high-speed hydraulic press. An attractive feature of the new press was its ability to do fast setups which was important because the plant produced many different products in low volumes. The new press cost $450,000, had an economic life of ten years, would reduce annual operating costs by $80,000, and would increase contribution to overhead and profits by $40,000 each year. The press would be depreciated over five years (using straight line depreciation). The plant had a tax rate of 40 percent. The discount rate was 12 percent, which was made up of a 10 percent risk-free rate and a 2 percent adjustment to reflect the risk in this type of project.

The investment satisfied strategic criteria S1 and S2. The press was an appropriate hard technology for a batch flow production system, and the ability to do fast setups would help provide higher levels of flexibility and innovativeness—which were the market qualifying and order winning outputs. The next step was to check criterion E1.

E1: How Safe Is the Investment?

The after-tax cash flows for the investment proposal are shown in Figure 12-2. The payback period is 4.2 years, that is, the total after-tax cash flow in the first 4.2 years is $450,000, which is the cost of the investment. If this payback exceeds the maximum allowed payback, the proposal is rejected and the assessment process stops. Otherwise, the investment is assessed to be safe, and the process continues to the next step, economic criterion E2.

E2: How Profitable Is the Investment?

The return on investment is

ROI = 90,000 / 450,000 = 20 percent

End of Year	Capital Costs	Depreciation[1]	Operating Cost Savings[2]	Contribution to Overhead and Profit[3]	After-Tax Cash Flow[4]
0	$450,000	--	--	--	--
1	--	$90,000	$80,000	$40,000	$108,000
2	--	90,000	80,000	40,000	108,000
3	--	90,000	80,000	40,000	108,000
4	--	90,000	80,000	40,000	108,000
5	--	90,000	80,000	40,000	108,000
6	--	--	80,000	40,000	72,000
7	--	--	80,000	40,000	72,000
8	--	--	80,000	40,000	72,000
9	--	--	80,000	40,000	72,000
10	--	--	80,000	40,000	72,000

Average = $90,000

Notes:

1. In each of the first five years there is $90,000 of depreciation. This reduces taxes by $90,000 × 40% = $36,000.
2. In each of the ten years there is a savings in operating cost of $80,000. Forty percent of this savings will be paid in taxes, leaving an after-tax cash flow of $80,000 × (1.0 − .4) = $48,000.
3. In each of the ten years there is an additional contribution to overhead and profit of $40,000 because of reduced scrap, increased production due to lower setup times, and so on. 40 percent of this contribution will be paid in taxes, leaving an after-tax cash flow of $40,000 × (1.0 − .4) = $24,000.
4. The after-tax cash flow is the sum of the after-tax saving due to depreciation, the after-tax operating cost saving, and the after-tax contribution to overhead and profit.

FIGURE 12–2

After-Tax Cash Flows For Situation 12.3

The internal rate of return is calculated from

$$
\begin{aligned}
450{,}000 = {} & 108{,}000/(1 + \text{IRR}) + \\
& 108{,}000/(1 + \text{IRR})^2 + \\
& \ldots + 108{,}000/(1 + \text{IRR})^5 + \\
& 72{,}000/(1 + \text{IRR})^6 + \\
& 72{,}000/(1 + \text{IRR})^7 + \\
& \ldots + 72{,}000/(1 + \text{IRR})^{10}
\end{aligned}
$$

which gives IRR = 17 percent. The net present value, when the discount rate is 12 percent, is

$$NPV = -450{,}000 + 108{,}000/(1 + 0.12) + 108{,}000/(1 + 0.12)^2 + \ldots + 108{,}000/(1 + 0.12)^5 + 72{,}000/(1 + 0.12)^6 + 72{,}000/(1 + 0.12)^7 + \ldots + 72{,}000/(1 + 0.12)^{10}$$

$$NPV = \$86{,}588$$

These profitability estimates (ROI = 20 percent, IRR = 17 percent, and NPV = $86,588) are average but not outstanding. They would satisfy most organizations' minimum requirements for profitability.

STEP 5: STRATEGIC CRITERION S3

What Effect Will the Investment Have on Other Manufacturing Levers? Is There Sufficient Manufacturing Capability to Make the Required Adjustments to the Levers?

The adjustments to the manufacturing levers needed to implement the proposed investment are scrutinized. An assessment is made of the organization's ability to make these adjustments. In general, an organization with a high level of manufacturing capability is better able to implement a difficult investment in a short time than an organization with a low level of manufacturing capability. The organization's track record for implementing new investments in manufacturing is also assessed. For example, what fraction of the potential benefits has the organization realized from past investments in manufacturing?

STEP 6: ECONOMIC CRITERION E3

How Robust Is the Investment?

The final step in the assessment process is a sensitivity analysis to determine how sensitive the profitability of the proposed investment is to the uncertainty in the important variables. For example, how certain are the estimated costs, the life of the investment, the size and timing of the future cash flows, and so on. Uncertainty that, when resolved, produces lower costs and larger, earlier cash flows is not a problem. What is a problem, however, is uncertainty that may result in an investment that would not satisfy economic criteria E1 and E2.

Three values—optimistic, pessimistic, and most likely—are estimated for each important variable in the investment proposal. The probability that each value will occur is also estimated. Measures of profitability such as NPV are then computed from these data. This is not an onerous task if modern spreadsheet software is used to do the calculations.

SITUATION 12.4

Stamping Plant Investment in New Equipment (Continued)

MANAGEMENT at the stamping plant realize that a considerable amount of uncertainty surrounds the savings estimated for the new press. Purchasing the new press has passed the first five checks. A sensitivity analysis is now undertaken to determine what effect changes in these estimates will have on the profitability of the proposed investment.

The estimates shown in Part 1 of Figure 12-3 are obtained for the optimistic, most likely, and pessimistic values of the operating cost savings and the contribution to overhead and profit. An NPV is computed for each combination of savings and contribution. Since there are nine combinations, Part 2 of the figure shows nine NPVs. Each NPV has a particular probability of occurrence. For example, if the optimistic annual savings of $90,000 is realized and the most likely annual contribution of $40,000 is achieved, then the NPV is $120,489 when the discount rate is 12 percent. The probability of this occurring is $0.2 \times 0.4 = 0.08$.

The NPVs are listed and plotted in Figure 12-3. The graph of the NPV (Part 3 of the figure) describes the robustness of the investment. For example, the probability that the investment will result in a loss, which is the probability that the NPV is less than zero, is 0.4. The largest possible loss is $82,919. The probability of above average profitability, say, NPV greater than $75,000, is 0.48, and the probability of high profitability, say, NPV greater than $100,000, is 0.26. In this way, management can assess how sensitive the profitability of the investment is to changes in the estimates of the important variables.

Part 1 Estimates for the Important Uncertain Variables

Variable		Optimistic	Most Likely	Pessimistic
Operating cost savings	Value Probability	$90,000 .2	$80,000 .4	$40,000 .4
Contribution to overhead and profit	Value Probability	$45,000 .3	$40,000 .4	$30,000 .3

Part 2 All Possible Values of Net Present Value

Combination		1.	2.	3.	4.	5.	6.	7.	8.	9.
Operating cost savings	Value Prob.	$90k .2	90k .2	90k .2	80k .4	80k .4	80k .4	40k .4	40k .4	40k .4
Contribution to overhead and profit	Value Prob.	45k .3	40k .4	30k .3	45k .3	40k .4	30k .3	45k .3	40k .4	30k .3
Net present value, NPV	Value Prob.	$137k .06	120k .08	87k .06	104k .12	87k .16	53k .12	-32k .12	-49k .16	-83k .12

Part 3 Net Present Values and Probabilities of Occurrence

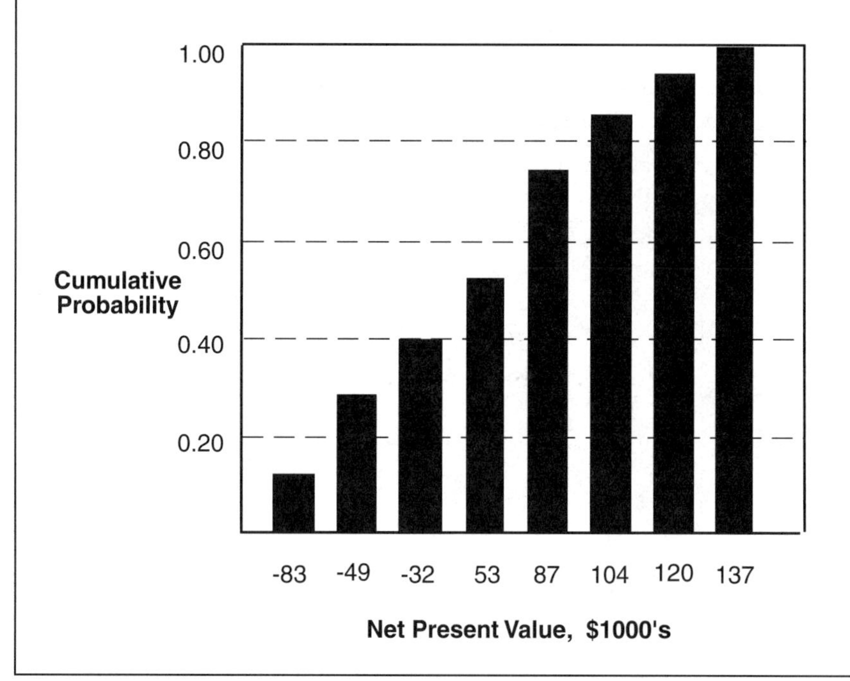

FIGURE 12–3
Using Sensitivity Analysis to Assess Economic Criterion E3

SITUATION 12.5

Redirecting Investment from Inventory to Hard Technology

RECALL the incident at the beginning of this chapter, in which a manufacturing manager wanted to retrofit a production process using money made available by reducing inventory. If the organization's cost of carrying inventory is 20 percent per year, then reducing inventories by \$750,000 will result in a cash saving of \$750,000 × 20% = \$150,000 each year. (The 20 percent cost of carrying inventory comprises the cost of capital, insurance, obsolescence, and any other relevant cash outflows incurred as a result of carrying inventory.) Suppose the organization has a 40 percent tax rate. Then the after-tax cash flow is \$150,000 × (1 − 0.4) = \$90,000 each year.

Retrofitting the production process will cost \$500,000. Suppose this investment is depreciated at the rate of \$100,000 each year for five years. Depreciation will result in an annual tax savings of \$100,000 × 0.4 = \$40,000 for each of the first five years. Finally, suppose that the new equipment in this project is expected to last eight years and the discount rate is 12 percent. The cash flows for the investment are summarized in Figure 12-4. The assessment of this investment follows.

FIGURE 12–4

After-Tax Cash Flows for Situation 12.5

End of Year	Capital Costs	Depreciation	Inventory Cost Savings	After-Tax Cash Flow
0	\$500,000	--	--	--
1	--	\$100,000	\$150,000	\$130,000
2	--	100,000	150,000	130,000
3	--	100,000	150,000	130,000
4	--	100,000	150,000	130,000
5	--	100,000	150,000	130,000
6	--	--	150,000	90,000
7	--	--	150,000	90,000
8	--	--	150,000	90,000
		--		
		--		

Average = \$115,000

Step 1: Is the Investment Appropriate for the Production System Used?

The production system was an equipment-paced line flow system, and the equipment that would be purchased in the proposal was appropriate for this production system.

Step 2: What Effect Will the Investment Have on the Market Qualifying and Order Winning Outputs?

The market qualifying outputs were cost and quality. Delivery was the order winning output. A large finished goods inventory was required to help provide delivery at an order winning level. Taking $750,000 out of this inventory would affect delivery time and delivery time reliability. Therefore, reducing inventory by $750,000 does not satisfy this criterion, and the proposal must be rejected. (The assessment process would stop here. However, we'll continue to see whether some insights into this proposal can be gleaned from the other four steps.)

Step 3: How Safe Is the Investment?

The payback is 3.85 years. Because the maximum payback period at this organization was four years, the investment is judged to be safe.

Step 4: How Profitable Is the Investment?

The average cash inflow is $115,000, and so ROI = 115,000/500,000 = 23 percent. IRR is computed from

$$500{,}000 = 130{,}000/(1 + IRR) + 130{,}000/(1 + IRR)^2 + \ldots + 130{,}000/(1 + IRR)^5 + 90{,}000/(1 + IRR)^6 + 90{,}000/(1 + IRR)^7 + 90{,}000/(1 + IRR)^8$$

which gives IRR = 17.4 percent. NPV is computed from

$$NPV = -500{,}000 + 130{,}000/(1 + 0.12) + 130{,}000/(1 + 0.12)^2 + \dots + 130{,}000/(1 + 0.12)^5 + 90{,}000/(1 + 0.12)^6 + 90{,}000/(1 + 0.12)^7 + 90{,}000/(1 + 0.12)^8$$

which gives NPV = $91,277. ROI is high, and IRR and NPV are average. The investment has average profitability.

Step 5: What Effect Will the Investment Have on Other Manufacturing Levers? Is There Sufficient Manufacturing Capability to Make the Required Adjustments to the Levers?

The new equipment is an adjustment to the process technology lever. Training will be needed for operators and maintenance personnel, which is an adjustment to the human resources lever. An adjustment must be made to the production planning and control lever (in the areas of scheduling, lot sizing, and so on) to ensure that delivery time does not increase when the $750,000 is taken out of finished goods inventory. Adjustments may also be needed to the sourcing and facilities levers. The organization's high level of manufacturing capability and track record at implementing new technologies suggest that it will be able to make all the adjustments needed.

Step 6: How Robust Is the Investment?

The manager may be a little optimistic in his estimate of the total cost of the project. There is a 50 percent chance that an additional $60,000 per year will be needed in the first two years for training, spare parts, and equipment rearrangement. Because the cost will be expensed, the after-tax cash outflow is $\$60{,}000 \times (1 - 0.4) = \$36{,}000$ in years 1 and 2. The resulting NPV is $30,467, a considerable drop from the earlier profitability estimate.

In summary, the investment proposal is rejected because it failed to pass step 2. As far as additional insights into the proposal are concerned, the proposal is safe but the profitability is low when possible additional costs to implement the investment are taken into account.

SUMMARY

Confusion and frustration often surround the process of evaluating investments in manufacturing. A systematic approach that takes account of strategic and economic concerns is needed. The approach presented in this chapter consists of a sequence of six checks or criteria, of which three are strategic and three are economic. The strategic criteria assess whether the proposed investment is consistent with the manufacturing strategy and, therefore, with the corporate and business unit strategies. The economic criteria gauge the safety, profitability, and robustness of the proposed investment. These economic criteria are well-known issues in modern financial management. Measures exist for each, but there is no single measure that combines all three. Most companies use multiple measures when they do capital budgeting, with IRR being the most popular, followed by payback and then NPV. See Kim, Crick, and Kim 1986; Shapiro 1990, (Chapter 6); and Horngren, Foster, and Datar 1994, (Chapters 20 and 21). Hayes and Wheelwright (1984, pp. 138–163) were first to translate these financial management issues into concepts that were easy to use in manufacturing strategy.

NOTES

1. Based on a problem from T. Hill, *Manufacturing Strategy*, Homewood, IL: Irwin, 1989, p. 89.

2. See page 139 of Hayes and Wheelwright 1984.

FURTHER READING

Hayes, R., and S. Wheelwright, *Restoring Our Competitive Advantage: Competing Through Manufacturing*, New York: John Wiley & Sons, pp. 138–163, 1984.

Horngren, C., G. Foster, and S. Datar, *Cost Accounting: A Managerial Approach*, 8th Edition, Englewood Cliffs, NJ: Prentice-Hall, 1994.

Kim, S., T. Crick, and S. Kim, "Do Executives Practice What Academics Preach?' *Management Accounting*, Vol. 68, No. 5, pp. 49–52, November 1986.

Krinsky, I., and J. Miltenburg, "An Alternate Method for the Justification of Advanced Manufacturing Technologies," *International Journal of Production Research*, Vol. 29, No. 5, pp. 997–1015, 1991.

Meredith, J., and N. Suresh, "Justification Techniques for Advanced Manufacturing Technologies," *International Journal of Production Research*, Vol. 24, No. 5, pp. 1043–1057, 1986.

Shapiro, A., *Modern Corporate Finance*, New York: Macmillan, 1990.

Part III

THE SEVEN PRODUCTION SYSTEMS

CHAPTER 13

THE JOB SHOP PRODUCTION SYSTEM

PRODUCTS AND VOLUMES

Most organizations have one or more plants or PWPs organized as job shop production systems. These general-purpose facilities produce low volumes of a wide variety of different products (see Figure 13-1). Once produced, there is no assurance that the same product will be produced again.

LAYOUT AND MATERIAL FLOW

In a job shop production system, departments and equipment are arranged in a functional layout; that is, equipment and processes of the same type are located in the same department. Employees work in one department only and are highly skilled on the equipment there. Because many different products are produced in low volumes, specialized equipment and tooling are not feasible, and general-purpose equipment and tooling are used. Most often the material flow through the job shop is different for each job. Work-in-process inventory is high. Delivery times are long because a great deal of material handling is required to move jobs from department to department, and jobs are set aside for lengthy periods waiting for busy equipment and operators to become available.

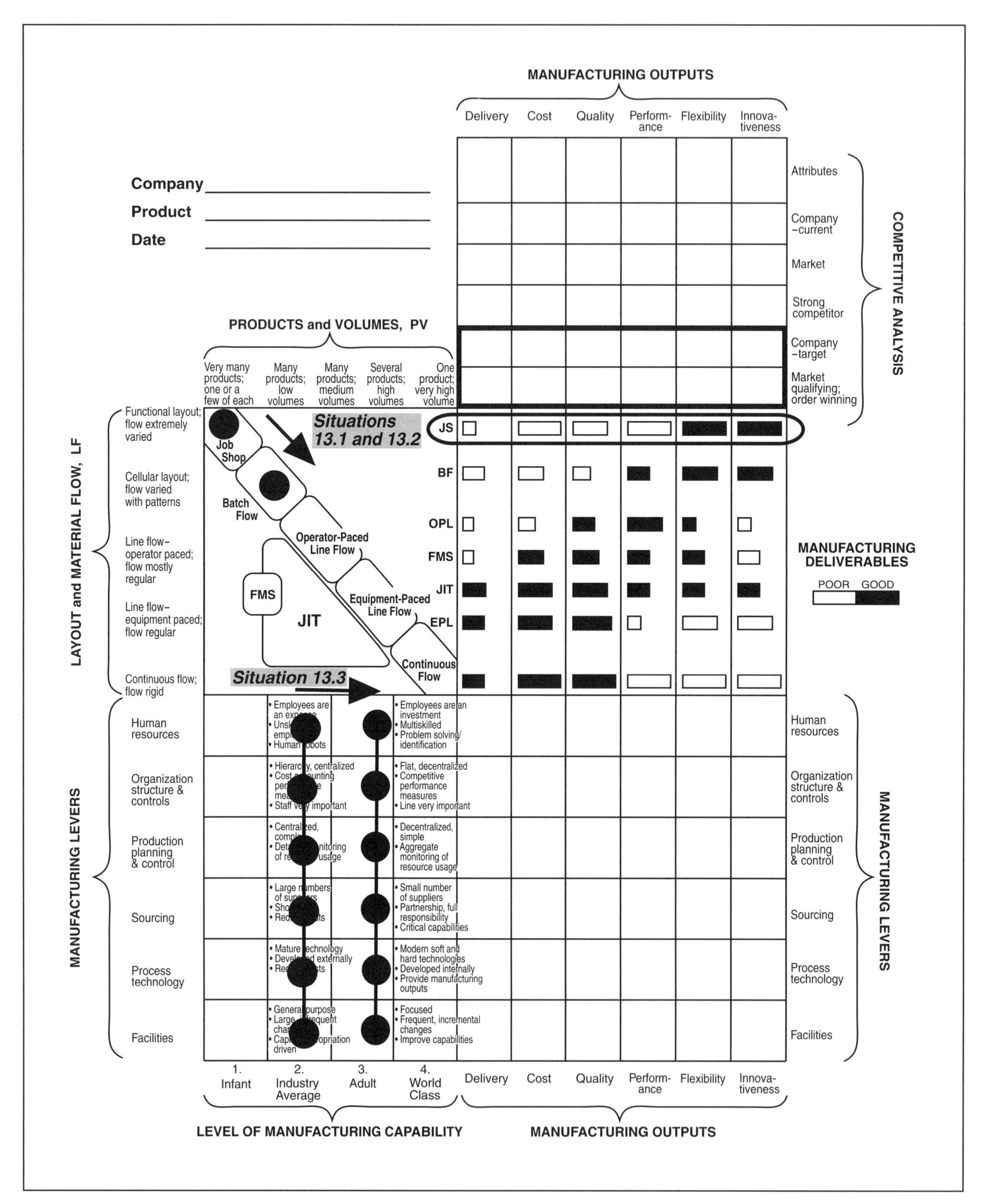

FIGURE 13–1
The Job Shop Production System

Competitive Advantage

Machine shops, welding shops, specialty clothing manufacturers, specialty electronics manufacturing, upscale restaurants, and hospitals are well-known examples of job shop production systems. Each provides great product variety and caters to individual customer demands. Job shop production systems are designed to provide high levels of flexibility and innovativeness, which they need to respond to their customers' demands for a wide variety of existing and new products. The job shop production system provides relatively low levels of the other manufacturing outputs—cost, quality, performance, and delivery (see Figure 13-1). This is not a handicap when competitors are also using job shop systems. However, if a competitor can manufacture the same products on a flexible manufacturing system (FMS), or line flow production system, the levels of the cost, quality, performance, and delivery outputs provided by the job shop system will be much too low. The conventional wisdom is to use the job shop production system only in situations where the volumes are too low and the differences between products are too great to permit batch flow or line flow systems.

In an attempt to raise the levels of the cost, quality, and delivery outputs, some manufacturers are changing their job shop production systems to batch flow, FMS, JIT, and operator-paced line flow systems. (Situations 13.1 and 13.2 describe such instances.) Another way to raise the levels of these outputs is to raise the level of the manufacturing capability of the job shop system. An example of this is the job shop production system at Standard Aero Ltd. of Winnipeg.[1] In January 1990, the company started an intense implementation of the total quality management (TQM) improvement approach (see Chapter 9) to raise its industry average level of manufacturing capability. Fifteen months later, the job shop's manufacturing capabilities had improved so much that the company won a $10 million contract to overhaul aircraft engine gearboxes for the U.S. military. The company's cost and delivery outputs were so much better than their competitors' that the Pentagon dispatched 13 senior officers to visit the company and investigate the bid. They were so impressed with Standard Aero's capabilities that the company was awarded the large contract.

Manufacturing Levers in the Job Shop Production System

Human Resources

Highly skilled employees operating general-purpose machines are the critical resource in the job shop production system. Usually there is little in-house training because operators learn their skills in apprenticeship programs outside the organization. Once they are trained, there is little need for the employees to improve these skills. Because it is difficult to find and keep these highly skilled employees, wages are high relative to other production systems. Incentive pay schemes are also used. Staffs are small. They concentrate on bidding for new work, working with customers to finalize the designs of new products, and expediting orders through the plant.

Organization Structure and Controls

Organizations with job shop production systems have organizational structures that are flat and organized by function. Staff departments are small and are less important than line departments. Operations are decentralized and entrepreneurial in nature so that they can respond quickly to customers' changing needs. Fixed costs such as buildings and equipment are low, and variable costs such as material and labor are high. Consequently, tight control is exercised over material and labor. Equipment is often idle, but labor is used efficiently. Line departments prefer to have a backlog of jobs so that they are always busy. This improves labor and equipment utilization but increases delivery time and work-in-process inventory. Quality is the responsibility of equipment operators and their immediate supervisors.

Sourcing

There is little vertical integration. Some purchased materials are stocked but most are purchased for specific customer orders. Because purchase orders are small and irregular, and many different suppliers are used, the job shop production system has little control over its suppliers.

Production Planning and Control

Orders are received through a process of competitive bidding. Production is make-to-order. Raw material inventory and finished goods inventory are small. Work-in-process inventory is large. Significant differences can exist between products, and products can be quite complex. Routings are developed for each product. The routings and product drawings accompany each order or job through the plant. Shop floor control, which consists of issuing dispatch lists to work centers listing the jobs to be produced with the deadlines, and input-output control to monitor the flow of orders and use of resources, are difficult. Expediting is used extensively to push important jobs through the plant. Scheduling employees is easy; scheduling equipment is not. Overtime is used when extra capacity is needed.

Maintenance is relatively easy. Breakdowns are not as disruptive as in other production systems because there are multiple units of each type of machine. When one machine breaks down, production is moved to another. The general-purpose machines are easier to maintain and repair than specialized, highly automated machines in the FMS and line flow production systems. Maintenance departments also develop considerable expertise in maintaining the multiple units of each type of machine.

Process Technology

There are no economies of scale in the job shop production system. Job shops are labor intensive. General-purpose machines with considerable flexibility are used. The machines tend to be old because technological change is slow. The job shop is a technology follower, not a technology leader. Setups are long because so many different products are produced. Run times are longer than they would be on more specialized machines. There are frequent capacity imbalances between departments because of the changing product mix on the shop floor. This causes bottlenecks and results in backup of jobs at various machines. The pace of production is slow because of the many delays, long setups, and slow machines. Quality is the responsibility of the machine operators. There is a large final inspection and test area, however, where the many different products are carefully checked and tested before shipment to customers.

Facilities

Facilities are small and general purpose. They are often old. For example, many job shop production systems are located in old multistory buildings where departments are located on different floors. Storage areas for raw materials, purchased components, and finished goods inventories are small. Work-in-process inventory is stored on the shop floor, which makes the job shop look crowded and somewhat disorganized.

Manufacturing Outputs Provided by the Job Shop Production System

Cost and Quality

Although the quality of the products produced in the job shop production system is satisfactory (otherwise the customer would not accept the product) it is difficult to meet tight specifications. This is because the equipment is general purpose, the tooling is low volume, and the volumes are too low to enable learning to occur (see Chapter 11). Similarly, although the cost is satisfactory (otherwise the customer would not buy the product) it is very difficult for the job shop to match the cost of a competitor that produces the same product in much higher volumes using specialized equipment in a line flow production system.

Performance

The job shop production system often produces products with a high level of performance. Compared to other production systems, however, it cannot provide the highest possible level of performance, year after year. The production volume of any particular product is so low that the manufacturer cannot afford the design engineering time required to design new, advanced features into the product, and the process engineering time needed to design new processes for producing these products.

Delivery

A job shop production system can provide on-time deliveries for its customers by expediting orders when necessary. However, expediting cannot be done all the time. The functional layout increases the distance a job travels and the amount of material handling required. The variety and number of jobs in the plant at any one time make scheduling difficult and create delays at bottleneck machines. It takes longer to do setups and run the required operations on general-purpose machines than it does on more specialized machines. All these factors increase the time required to complete a job.

Flexibility and Innovativeness

The job shop production system is designed to produce a wide variety of products in very low volumes. Job shops specialize in producing customized products. Equipment and tooling are general purpose and equipment operators are highly skilled. Consequently, it is relatively easy to change product mix and volumes, to make design changes, and introduce new products. The job shop production system provides the highest possible levels of flexibility and innovativeness.

The following examples describe situations when organizations with job shop production systems sought to provide higher levels of the manufacturing outputs. In Situations 13.1 and 13.2, the organizations implemented group technology principles to change their job shop systems to batch flow systems (see Chapter 14). In Situation 13.3, the organization did not change its production system, rather it made numerous improvements to each manufacturing lever, thereby raising the level of manufacturing capability and achieving higher levels of all the manufacturing outputs.

Situation 13.1

Heavy Equipment Manufacturer Changes Job Shop to Batch Flow

MECHANICAL parts for medium- and heavy-duty trucks were manufactured in a plant using a job shop production system. The functional layout is shown in Figure 13-2. One of the many different parts produced by the job shop was a front end spindle. The material flow for this part was particularly complex because the part was complicated. More than 12 different spindles were produced, but the total volume was too low to justify a dedicated line flow production system for the part.

FIGURE 13–2

Moving from Job Shop to Batch Flow in Situation 13.1

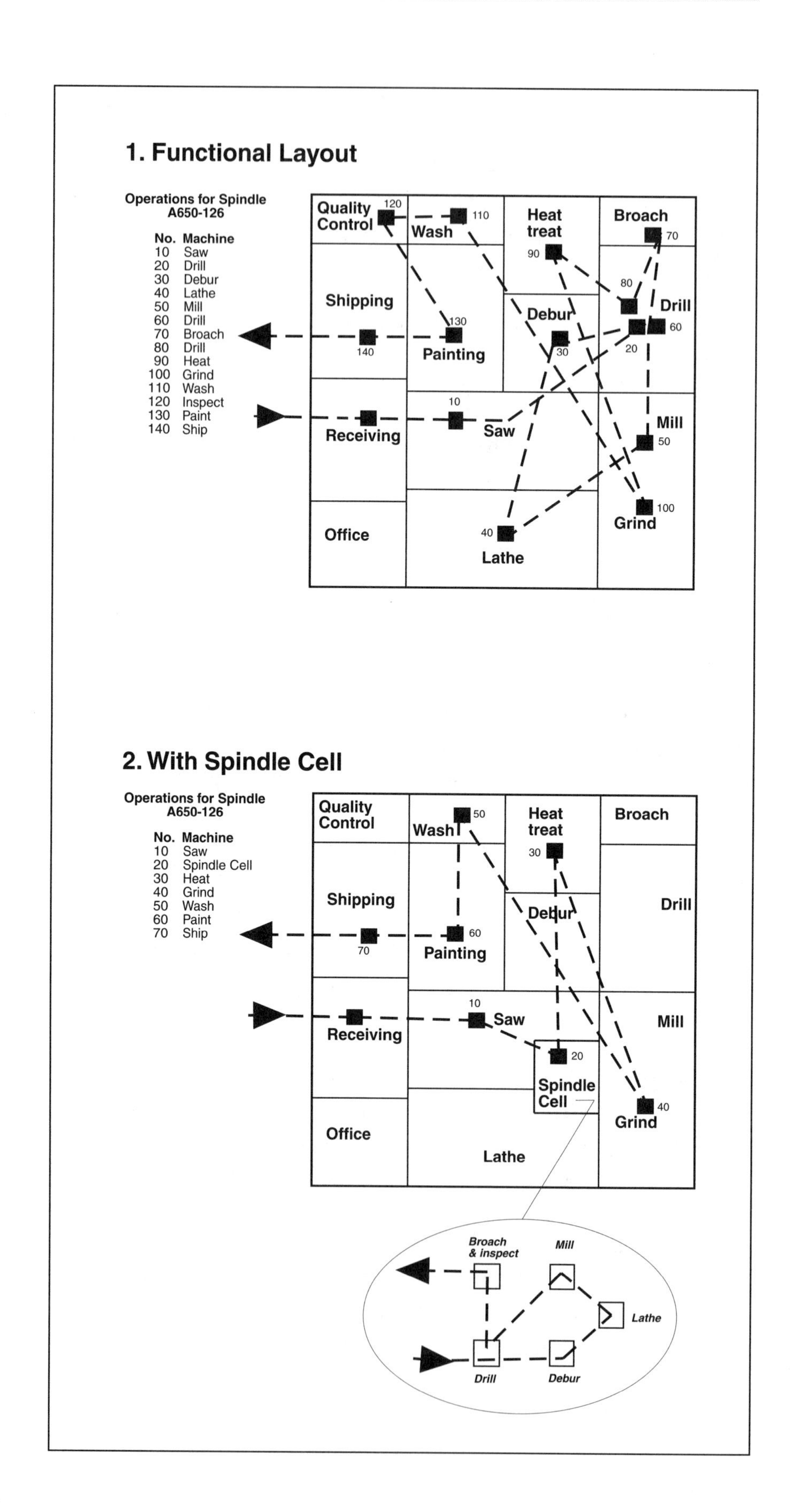

The plant was under pressure from its customers to reduce the cost of all its parts and to improve quality and delivery. The plant applied some group technology concepts and found, among other things, that a part family consisting of front end spindles and other similar parts had a sufficiently high total volume that it could be produced in a dedicated manufacturing cell. A drill, debur, lathe, mill, and broach were moved into the new cell, with some test equipment. (See Figure 13-2.) The spindle cell parts were sufficiently different in terms of setups, tooling, and operations that the best way to produce them was in batches. Consequently, space was left around each machine where containers of parts waiting to be processed could be placed. The result was a batch flow production system for the spindle parts. The new production system provided better levels of cost, quality, and delivery for these parts than did the old job shop production system.

SITUATION 13.2

Applying Group Technology at John Deere

IN THE MID-1980s John Deere began applying group technology (also called "cellular manufacturing") principles to its many job shop production systems to convert them to batch flow and line flow systems. The results were satisfying. In one plant, for example, moving machines into manufacturing cells, assigning part families to the cells, and tailoring the machines and tooling in each cell for the part family resulted in a 25 percent reduction in the number of machines required in the plant, a 70 percent reduction in the number of departments handling an average part, and reductions in setup times and material handling costs. These and other improvements raised John Deere's manufacturing capability and enabled it to provide higher levels of the manufacturing outputs than its competitors. This helped the company withstand the slump in the farm implements market during the late 1980s and early 1990s.

SITUATION 13.3

Improving the Job Shop Production System at CWM Industries

CWM COMPANY was a multinational producer and marketer of small-volume, heavy-duty mechanical products for use in forest harvesting, agriculture, and construction equipment. The company's products were designed to customer specifications and produced in small lots. Prices ranged from several hundred dollars to over $10,000. CWM was concerned about the number of late deliveries, long delivery times, and the high level of work-in-process inventory.

The company had four plants in the United States and Canada. About two-thirds of the floor space in each plant was devoted to general-purpose machine shops. The machine shop in the largest plant was a job shop production system, with 320 general-purpose machine centers arranged in a functional layout. A typical part was routed through 10 to 15 machine centers during the fabrication process. When a part reached a machine center, it waited in a queue until the machine became available. The operator set the machine so that the part could be produced according to the specifications on the drawing that accompanied it. The remaining one-third of the plant was used for assembly and testing. A typical product was assembled from about 200 parts and assemblies. The lot size was determined by the production control department and was usually set equal to the number of units in the customer order.

The company was sure that the job shop production system was best for it. After visiting plants in other industries with the same production system, however, the management realized that they had an industry average level of manufacturing capability, which they would need to raise to at least an adult level to solve their delivery and inventory problems and provide the manufacturing outputs at the required levels. Improvements to each manufacturing lever would be necessary. Some of the improvements they set out to make included the following:

1. Improve production planning and control, control the release of new orders to avoid over-

loading the shop, and use input/output control at each important machine center to ensure that schedules are met.

2. Increase the number of common components, produce these to a forecast, and store them in inventory.
3. Work with important suppliers to improve delivery time and delivery time reliability, and book capacity at suppliers' plants ahead of time and release orders against this capacity at the latest possible moment.
4. When possible, standardize machine centers and eliminate unnecessary machines.
5. Reduce setup times on the most heavily used machines and schedule and control these machines carefully.
6. Begin a limited program of training employees to operate more than one machine.

Notes

1. See T. Wakefield, "No Pain, No Gain," *Canadian Business*, pp. 50–54, January 1993.

CHAPTER 14

The Batch Flow Production System

Products and Volumes

The batch flow production system produces low volumes of many different products (see Figure 14-1) for one of three reasons: 1) Customers place one-time orders for small quantities, 2) customers prefer to place small repeat orders from time to time rather than one large order, and 3) the organization prefers to accumulate orders from different customers for the same or similar products into a batch and produce them together. Regardless of the reason, production of products is make-to-order. Common parts, when they exist, can be make-to-stock.

Layout and Material Flow

Functional and cellular layouts are used in the batch flow production system. In a functional layout, equipment of the same type is located in the same department. In a cellular layout, different types of equipment are located in the same department so that all the operations required to produce any product within a product family can be performed in the department or cell. Group technology principles are used to determine product families and equipment cells.

The material flow in the batch flow production system is varied, but there are patterns when products are produced in cells.

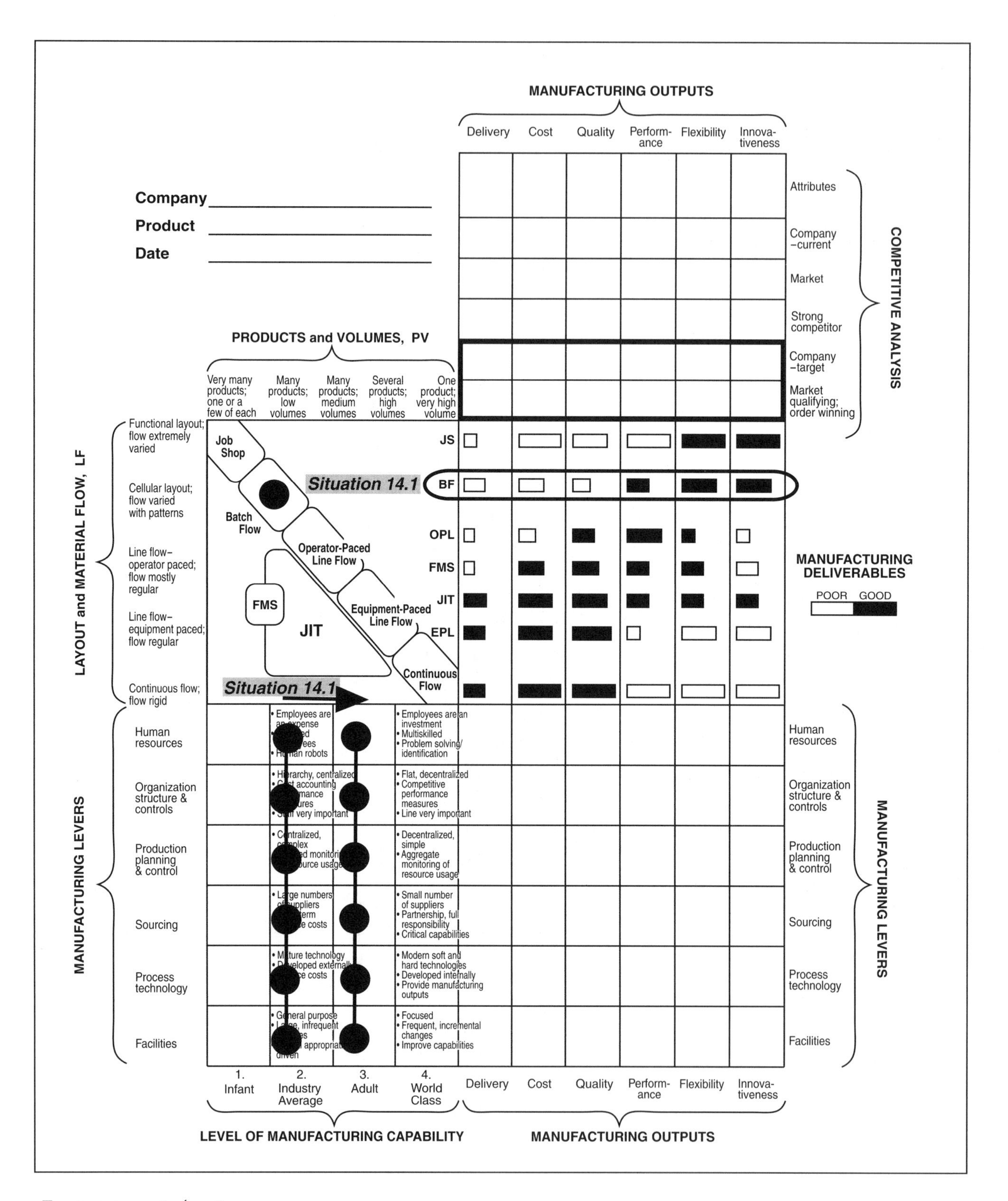

FIGURE 14–1

The Batch Flow Production System

When the functional layout is used, the material flow is irregular and considerable material handling is needed to move the batches from department to department. Once moved, the batches endure long delays waiting for equipment and operators to become available. When the cellular layout is used, the material flow is less varied. Some material handling is needed to move small batches of different products among the machines in the cells, and delays, though shorter, still occur.

Competitive Advantage

The batch flow production system produces many different products in low volumes. Consequently, it is easy to change volumes, introduce new products, or make design changes to existing products. The production system is designed to provide high levels of flexibility and innovativeness, which are necessary to respond to customers' demands for a wide variety of products.

The volume of a product in the batch flow system is higher than it would be in the job shop. Fixed costs, such as the cost of setups, can thus be spread over more units and learning can occur—that is, improvements can be made (see Chapter 9). The results are better levels of cost and quality than in the job shop production system. However, the levels are not as good as they are for the line flow systems. The batch flow system provides a higher level of performance than the job shop production system. Because the volume of any particular product or product family is higher, the organization can assign more engineering resources to improve the products' features by making design changes to the products and improvements to the manufacturing process. The volume is not as high as it is for the line flow production systems, however, so the organization cannot afford to assign as many resources as these systems can. The conventional wisdom is to use the batch flow production system when producing many different products in volumes that are too low for an operator-paced line flow production system.

An organization has two choices when it is not satisfied with the levels of the manufacturing outputs provided by its batch flow production system. It can make improvements to increase the level of manufacturing capability (see Chapter 9), or it can change the system to a just-in-time production system (see Chapter 17).

Manufacturing Levers in the Batch Flow Production System

Human Resources

Employees are assigned to one department or a manufacturing cell and are trained to operate all the machines there. Incentive pay schemes are used. Staffs are small. They concentrate on bidding for new work, working with customers to finalize the designs of new products, and expediting orders through the plant.

Organization Structure and Controls

Like the job shop production system, operations are decentralized and entrepreneurial so that they can respond quickly to changing customer needs. The organizational structure is flat. The staff departments are small and are less important than the line departments. There is no quality department. Quality is the responsibility of machine operators and their supervisors.

Sourcing

There is little vertical integration. Some purchased materials are stocked, but most are purchased for specific customer orders. Because purchase orders are small and irregular, and many different suppliers are used, manufacturers using the batch flow production system have little control over their suppliers.

Production Planning and Control

Traditional MRP II systems are used to plan and control all production activities. Orders are usually received through competitive bidding. Production of products is make-to-order; production of common components is make-to-stock. Raw material and finished goods inventories are small. Work-in-process inventory is large.

Management in a batch flow production system is more concerned about equipment utilization than management would be in a job shop system. To increase utilization, orders for the same product or similar products are batched and manufactured together. This reduces setup costs and generates efficiencies. Management is also concerned that orders are completed on time. Like the job shop production system, products back up waiting for equipment to become available. When this threatens to make an order late, the order is expedited through the plant.

Process Technology

Equipment and tooling are mostly general purpose, but there is some specialization when the volumes of families of similar parts permit manufacturing cells to be formed. Setup times are long because of the general-purpose equipment and tooling. The number of setups is high because of the large number of different products produced. Batch flow production systems are labor intensive. Machines tend to be old and the rate of technological change is slow. Like the job shop production system, the batch flow system is a technology follower, not a technology leader.

Facilities

Facilities in a batch flow production system tend to be larger than in a job shop system. Linkages between departments are loose. When one department completes its operations on a batch of parts, the batch is moved to the next department. There are frequent capacity imbalances between departments because of the changing product mix in the facility. The results are bottlenecks and buildups of work-in-process inventory in different departments at different times.

Manufacturing Outputs Provided by the Batch Flow Production System

Cost and Quality

Cost is more important in the batch flow production system than it is in the job shop system because 1) the number of units of each product is higher, so even a small reduction in the cost to produce a product can generate a large total saving, and 2) some customer orders are repeated from time to time. Cost can be an important factor in these customers' decisions about where to place their orders.

Because a small volume of each product is produced, there is an opportunity for learning to occur and improvements to be made. Fixed costs, such as the cost of setups, can be spread over more units. This results in improved levels of cost and quality compared to the job shop production system. Nevertheless it is difficult for a manufacturer with a batch flow system to match the cost and quality of a competitor who produces the same product in a much higher volume on special-purpose equipment in a line flow production system (see Figure 14-1).

Performance

The batch flow production system provides a higher level of performance than the job shop system because the volume of any particular product or product family is higher. This permits the organization to dedicate some modest engineering resources to work with customers to design new, advanced features into the products, and to design better processes and tooling for manufacturing the products. The volume is not nearly as high, however, as it is for the line flow production systems, so some of these systems will provide higher levels of performance.

Delivery

Because many different products are produced simultaneously, products will often compete for time on the same equipment. The pace of production is slow because of the build-up of work-in-process inventory at workstations, the long setup times, and the slow general-purpose machines. Scheduling is difficult because of the large number of orders. The result is a long delivery time and a low delivery time reliability. Fast deliveries can sometimes be provided by expediting orders.

Flexibility and Innovativeness

The batch flow production system can provide high levels of flexibility and innovativeness because of its general-purpose equipment and facilities and its skilled employees. Consequently, it is easy to change product mix and volumes, make design changes, and introduce new products.

Situation 14.1

Farm Implements Manufacturer Improves Batch Flow System

THE CJC plant produced four lines of farm implements—plows, seed drills, loaders, and spreaders. Sales in 1990 exceeded $350 million. The forge department, machine shop, weld shop, paint shop, and sub-assembly and final assembly departments employed 1,200 people. The plant fabricated 6,000 different parts and purchased 12,000 parts. Almost all fabricated parts had low volumes and were produced in batches. Some had different processing requirements, while others had similar requirements. The former were routed through departments organized by function. The latter were grouped into families, and when the volumes in the families were high enough, manufacturing cells were organized to produce them. The first operation for most fabricated parts occurred in the forge department, where steel plates and bars were sheared, heated, and formed into intricate shapes. A heat treat operation followed, and then the parts were sent to a machine shop for machining, drilling, and grinding. Some parts then went to the welding department for welding operations; others went immediately to a paint shop where they were cleaned and painted. From there, parts went to a subassembly department where they were assembled with other parts. Final products were produced on one of the final assembly lines, after which they were tested, packaged, and shipped to customers.

Job shop and batch flow production systems were used in the forge department, machine shop, paint shop, and welding shop. Attempts were under way to change the job shop production systems to batch flow systems, and to improve the batch flow systems so that higher levels of the manufacturing outputs would be provided. (See Figure 14-1.) This was not easy because of the size of the departments. The forge department had 100 workstations, the machine shop had 130 machine tools, and the welding shop had 40 workstations. Operator-paced line flow production systems were used in the subassembly and final assembly departments. A modern MRP II system was used to plan and control the production activities in all departments.

In the discussion that follows, the manufacturing levers in the batch flow production systems in the forge, machine, and welding shops are examined.

Human Resources at CJC

There were 120 job classifications for production operators in the three departments. Few operators were trained to operate more than two pieces of equipment in their department. The plant was unionized and jobs were assigned on the basis of seniority. Material handlers were responsible for moving material among workcenters. Equipment maintenance and repair were done by the maintenance department. Production operators were responsible for the quality of the parts they produced. No statistical quality control procedures were used. A small engineering department comprising industrial, process, plant, and product engineers was responsible for providing technical support for processes and products.

Inventories were used whenever possible to smooth the load on the plant. However, the demand for farm implements was seasonal, so overtime and layoffs were also used to bring production into line with the seasonal customer demands.

Industrial engineers were responsible for setting labor standards. The main purpose of standards was to support an incentive wage scheme, which determined compensation for the production operators. The labor standard was the time an operator was allowed to do a setup or complete an operation for a part on a machine. Because there were 6,000 fabricated parts and each part required an average of four operations in the forge department and machine shop, there were more than 24,000 labor standards in the two departments. Operator efficiencies of 125 percent, which meant that operators were able to produce 25 percent more than that calculated from the labor standards, were common. It was becoming increasingly difficult for the industrial engineers to keep the labor standards up-to-date. There were three reasons for the difficulty:

1. Each time an improvement was made, several labor standards had to be revised. As the pace

of making improvements increased, the backlog of standards needing revision increased. As soon as a standard was revised, a new improvement made it obsolete.

2. Industrial engineers were involved in identifying and implementing improvements, and often had little time for updating labor standards.
3. The incentive wage scheme based on labor standards did not work well in the manufacturing cells. The machines were more specialized and operators could run more than one machine at a time. Incentives based on individual labor standards disrupted teamwork in the cells.

CJC needed to make several adjustments to this lever. The incentive wage scheme had to be changed (more on this later), the number of job classifications reduced, and the job descriptions changed. CJC was discussing these and other changes with the union.

Organization Structure and Controls at CJC

Each department in the plant was a cost center. Departments were evaluated on their ability to keep actual costs below budgeted costs (which meant that machine utilizations and labor efficiencies had to be kept high) and their ability to complete orders on time. Supervisors and expediters spent a great deal of time rushing orders through workstations so that orders could be completed on time and workstations would not be forced down because of a shortage of parts.

Staff groups were small. They did production planning and control, and managed the information systems between the plant and its suppliers, and the plant and the sales groups. This was not an easy job because of the large number of parts and products manufactured at the plant.

Sourcing at CJC

In the past CJC had a high degree of vertical integration. This was about to change because the company recently decided to outsource many of its very low volume parts. Some parts were outsourced to existing suppliers. In many cases, however, they did not have the capability to manufacture the parts and so new suppliers had to be found. The plant did not have close ties or much control over its suppliers because the amount of business it had with any one supplier was low. Suppliers were currently selected on the basis of lowest cost from those who had the capability to produce the part. Because of the decision to purchase more from outside vendors, CJC would have to develop more expertise in managing its suppliers.

Production Planning and Control at CJC

The sales group developed a 12-month forecast for final products. The forecast was updated monthly to account for changes in the market. Capacity planning decisions concerning number of employees, levels of inventories, and amount of overtime were made on the basis of the forecast. An MRP II system released manufacturing orders to the plant and purchase orders to suppliers at the beginning of each week. Orders for final products usually represented actual customer orders. Sometimes orders for final products were for inventory—when it was necessary to anticipate seasonal demands. Common parts were produced for inventory. This reduced the lead time for final products and helped to smooth the workload in the plant.

Each order was released the length of time equal to its lead time before it was needed. Lead times were difficult to predict because they depended on many different factors, including the number of units to be produced, the setup times and run times at the workstations where the part would be produced, the workload at the workstations, equipment and tooling problems, material handling resources, and operator availability. Because there were more than 2,000 open orders on the plant floor at any one time, it was difficult to monitor and control the flow of material. A great deal of expediting was done to speed orders through the plant and avoid missing promised delivery dates, and to keep key workstations busy. Depending on which parts were produced, different workstations became bottlenecks, which required that overtime be scheduled to create needed capacity.

In the forge department, machine shop, and welding shop, each part was produced in a lot size equal to one month of requirements. Smaller lot sizes were desirable to reduce work-in-process inventory and lead times, but the setup times at most workstations were too long to permit smaller lot sizes to be produced economically. Moderate inventories of raw materials and purchased parts were maintained. Work-in-process inventory was large. Finished goods inventories were small unless products were produced early to meet large upcoming seasonal demands. The seasonal demand for some products created a great deal of uncertainty in the timing and quantity of purchased parts and materials.

The incentive wage scheme created some problems for production planning and control because it made it difficult to move employees from one workstation to another in response to changes in the workload. A new computerized system was implemented to remedy this. Under the new system, each employee would have a picture identification card with their employee number bar code on the back. Manufacturing orders would have the order number and operation information bar codes on an order card that traveled with the order. Reader stations would be located throughout the plant. Stations would consist of a bar-code reader and a simple keyboard for entering information about number of units completed, scrap units, and so on. When an operation was completed, the employee would go to a reader station, where the bar-code reader would read the employee number on the identification card, and the manufacturing order and operation numbers from the manufacturing order card. The employee would then be prompted for the number of units completed and the number of units scrapped. This would complete the data entry and the employee could return to the same workstation or go to a different one and proceed to the next order. The data would be sent to a central computer where payroll information would be computed, labor efficiency and workstation utilization would be calculated, the current status of all manufacturing orders would be monitored, and the level and location of work-in-process inventory would be tracked.

Process Technology at CJC

There were many different process technologies in the plant. The forge department had saws and shears for cutting a variety of steel bars, tubes, and sheets. There were small and large drop hammers, presses and wheelabraters, and different types of heat treating and stress relieving equipment. The machine shop had many types and sizes of drills, lathes, mills, and grinders. The welding department had many varieties of mig and arc welders, and manual and robotic welders. There were various metal cleaning and painting facilities. The technologies were traditional, in spite of their numbers. Machines and tooling were general-purpose—designed to produce many different parts—and the setup times were long.

All these processes were required to complete numerous operations on the thousands of different parts. No one technology was more important than the others. This, and the fact that the engineering staff group was small and very busy, made it impossible to remain current with new developments in process technologies. When new process equipment was acquired, it was difficult to realize all the potential benefits because the team implementing the new technology did not have the time or resources to make changes, which were necessary to exploit the capabilities of the new technology, to the processes upstream and downstream of the new equipment.

There was a proliferation of parts, machines, and tooling. Many machines were used only a few times each year when particular products were produced. Communications between the plant and the product design group, which was located hundreds of miles away, needed improvement. The plant had just started standardization and design for manufacturability programs. There were no formal quality control programs. Sometimes this resulted in batches of parts having to be reworked to correct a defect that was not detected in time. There was incoming inspection for parts received from suppliers, but there was no supplier certification program to ensure that suppliers were capable of delivering high-quality parts on time.

Facilities at CJC

The CJC plant was not large. However, it was a complex plant in terms of the many different parts produced and processes used. CJC reduced this complexity by decreasing the vertical integration (purchasing more parts from outside vendors) and by standardizing where possible. Each department in the plant was medium in size—too small to achieve significant economies of scale. There were loose links between departments; when the operations were completed in one department, the parts were sent to the next.

Changing Production Systems

There are many organizations like CJC, who make improvements to their batch flow production systems to increase their manufacturing capability. The goal is to raise the levels of the manufacturing outputs to meet the rising expectations of customers. Other organizations with batch flow systems are reaching for the same goal by changing their production systems to operator-paced or just-in-time production systems. (See Situation 3.2 in Chapter 3.) Regardless of what improvements and changes are made, the higher the existing level of manufacturing capability, the easier they will be to make (see Chapter 5).

Some organizations with job shop and batch flow production systems face an additional problem because these organizations are sometimes managed by entrepreneurs who have been with the organization from the beginning and may even have started the business. The tenacity and energy of these managers—characteristics necessary to build the business—can become obstacles to change as new conditions create the need for different production systems. Production systems further down the PV–LF matrix require management processes that are more objective and systematic and less hands-on than these entrepreneurial managers are used to. This problem must be taken into account when the manufacturing strategy and the implementation plan are developed.

CHAPTER 15

THE FLEXIBLE MANUFACTURING SYSTEM

Definitions of what constitutes a flexible manufacturing system (FMS) abound. Manufacturers of robots, automatic guided vehicles, and CNC machines have their own definitions. In this book, we define an FMS as a group of CNC machines linked together by an automated material handling system. All activities are integrated by a supervisory computer. An FMS can produce any product from a large product family in random order. An FMS can run unattended for long periods. FMSs range in size from small—five to ten machines, robots, and automated guided vehicles—to extremely large (see Figure 15-1).

SITUATION 15.1

An FMS Production System at Fanuc

THE FUJITSU FANUC ROBOT FACTORY is a well-known FMS production system. It produces 100 robots, 75 small machining centers, and 75 wire-cut electrodischarge machines each month. The factory's machining department produces small quantities of many different parts used in these products. Twenty-nine machining cells make up the department. Each cell consists of one or more CNC machines, with robots or pallet changers to load and unload them. Parts are fixed to pallets, and automatic guided vehicles move pallets from cell to cell and in and out of automated storage and retrieval warehouses. Nineteen operators run the machining department during the day. At night, the cells run unattended and a single operator watches from a control room. Machine availability is close to 100 percent, and machine utilization averages about 67 percent over 24 hours.

FIGURE 15–1

An FMS Production System

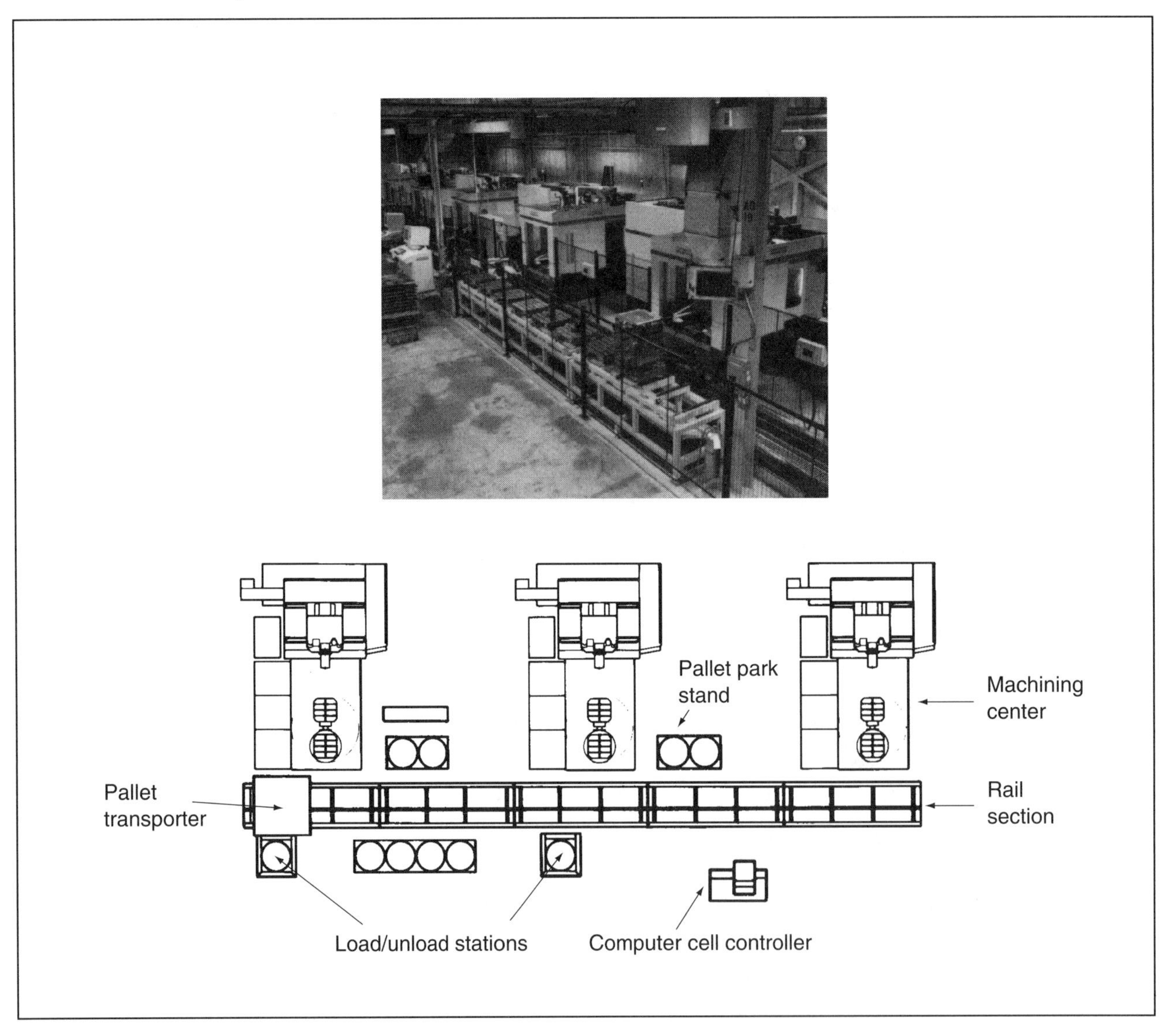

Source: Cincinnati Milacron 1994. With permission.

Products and Volumes

The FMS production system produces low volumes of many different products (see Figure 15-2). Production of products is make-to-order and in small batches because customers either place one-time orders for small quantities or place small repeat orders from time to time rather than one large order.

Layout and Material Flow

An FMS is a special line flow production system because the layout is cellular but the material flow is a line flow. In a cellular layout, different types of equipment and processes are located in the same department or cell, so that all the operations required to produce any part within a part family can be performed within the cell. Group technology principles are used to identify part families and equipment cells. Automated material handling—conveyors, robots, and automated guided vehicles—move the material through the FMS as though it were a line flow production system. Robots load material into the CNC machines, where automatic tool changers change tools, computers retrieve programs, and the machines quickly complete the required operations. Thus, low volumes of a wide variety of products are produced.

Competitive Advantage

There are three reasons why manufacturers use FMS production systems.

1. Manufacturing Outputs

The FMS production system is an alternative to the batch flow system and, to a lesser extent, the job shop system. All produce many different products in low volumes. However, the FMS provides the highest levels of the quality and cost outputs. Thus, the conventional wisdom is to use an FMS rather than a batch flow production system when the quality level provided by a batch flow production system is too low. The high level of quality in the FMS is provided by sophisticated, flexible equipment and tooling that, despite being general purpose, can produce products with very tight specifications. For some products, such as dies, fixtures, and parts for the aerospace

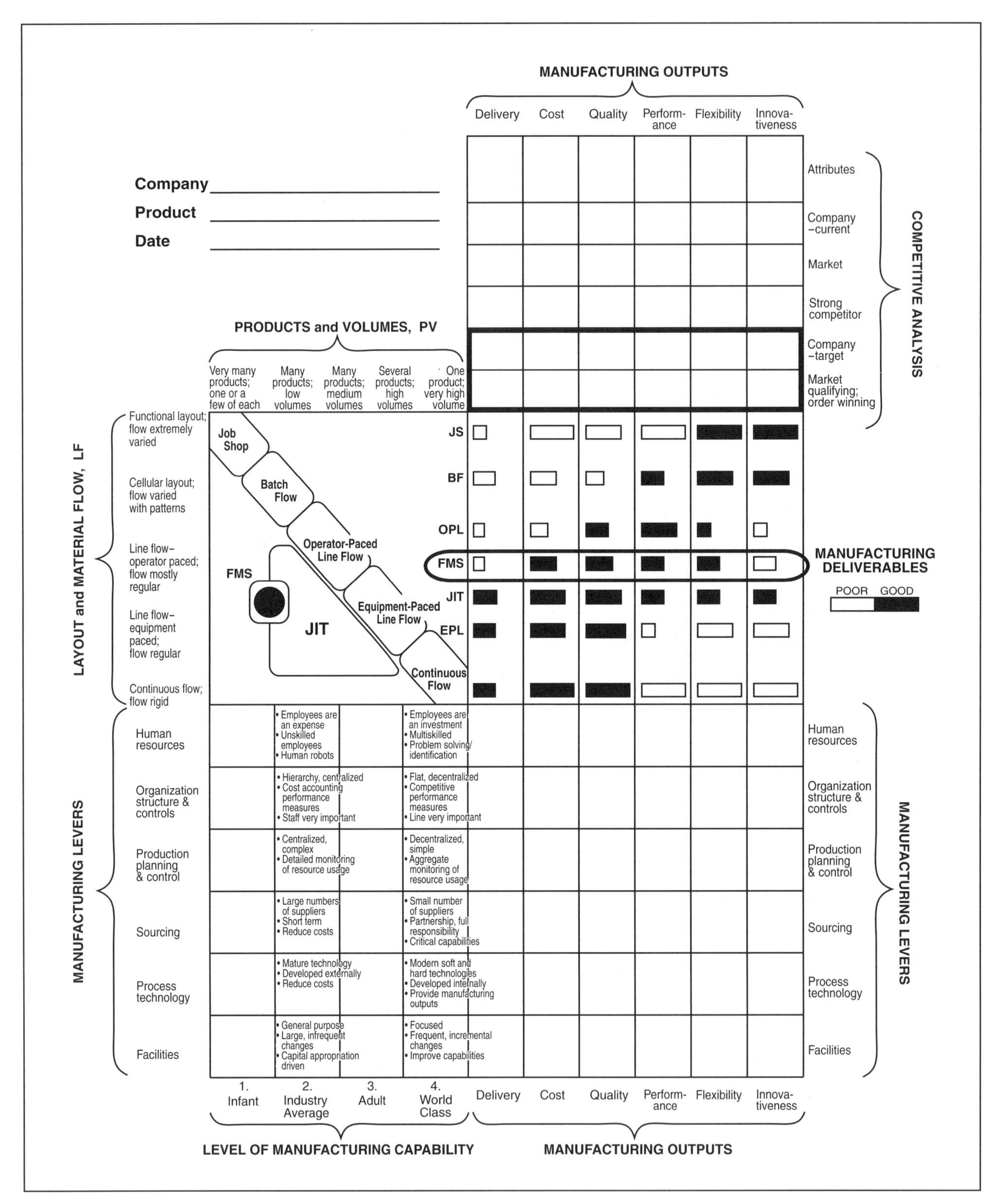

FIGURE 15–2
The FMS Production System

industry, the specifications are so tight that the products can be produced only on the type of computer-controlled equipment used in an FMS; a batch flow production system cannot be used.

When an FMS is utilized properly, the product cost will be low. This does not require 100 percent utilization. The FMS in Situation 15.1 ran at 67 percent utilization, and the FMS in Situation 15.3 was designed so that an 80 percent utilization would give products with a low cost. Because an FMS is a line flow production system, delivery time and delivery time reliability are also good. An FMS provides a high level of flexibility, with a somewhat lower level of innovativeness than a batch flow system. An FMS is designed to produce many different products in low volumes, so it is relatively easy to change products and volumes. It is more difficult to introduce new products and make design changes to existing products because new pallets, fixtures, and computer programs are required. However, an FMS provides a higher level of innovativeness than traditional line flow production systems.

2. Fast Changeovers

An FMS can be used when there are too many different parts and the setup times are too high for conventional machines in a conventional manufacturing cell. (See Situation 10.5 in Chapter 10.) Batch flow and operator-paced line flow production systems sometimes assign too many parts to a manufacturing cell to increase machine utilizations. The differences between parts may be small, but they are often large enough to make setups troublesome. Too much production time is lost doing setups. Setups are faster on the flexible machines in the FMS, so less production time is lost.

3. Special Expertise

A manufacturer may wish to gain technical and managerial expertise with flexible manufacturing systems as an investment for the future or as part of a product development program. Companies like John Deere and General Motors want expertise with the technologies that constitute the FMS so that they are in a position to exploit these technologies when their worth becomes evident. Companies like Cincinnati Milacron, Fanuc, and Allen-Bradley produce the machine tools, robots, and computer controllers used in an FMS. These companies use FMS production systems, rather than simpler batch flow or traditional line flow production systems, so that they can learn more about how their products work in an FMS. This helps them to improve their own products and provides a valuable showcase of their products for marketing purposes.

Manufacturing Levers in the FMS Production System

Human Resources

Flexible CNC machines, robots, and automated guided vehicles perform all the operations in the FMS production system. Only a few well-trained operators are needed to perform support tasks such as placing parts on pallets, putting parts in automatic material feeders, loading tools into tool carousels, and adding cooling fluids for the cutting tools. Maintenance is an important activity. Maintenance personnel are responsible for preventive maintenance, maintaining spare parts, and making repairs during breakdowns. The equipment in an FMS has a high level of mechanical and electronic complexity and is difficult to maintain. Maintenance personnel must be highly skilled and well trained. All employees are paid an hourly rate; no incentive pay scheme is used.

Organization Structure and Controls

Although the FMS production system can be a profit center, it is usually a cost center. Line departments are small because most of the work is done by machines. Numerous staff departments are required for activities such as obtaining new orders, production planning and control, writing computer programs for the CNC machines, designing new pallets and fixtures, and managing quality control and maintenance activities. A quality control department is responsible for all aspects of quality in the production system. This includes statistical process control activities in the production process, quality of purchased materials, and projects for improving quality.

Sourcing

There is little vertical integration. Some purchased materials are stocked, but most are purchased for specific customer orders. Because purchase orders are small and irregular, and because many different suppliers are used, the FMS production system has little control over its suppliers.

The machines, fixtures, and tooling in an FMS are precise. To work properly, they require materials with very tight tolerances. Consequently, the quality of materials received from suppliers must be high. Suppliers are evaluated on their ability to provide high levels of quality, short delivery times, and low cost.

Production Planning and Control

Orders are received from external customers through a process of competitive bidding or from internal customers through an allocation process. Production is make-to-order. Raw material and finished goods inventories are small. Work-in-process inventory is not large. Products can be complex, and there can be significant differences between products. Process plans, which include computer programs for each operation at each workstation visited by a product, are developed. This is a costly and time-consuming activity.

Elaborate computerized information systems are used to support a scheduling system, a control system, and other management systems. Scheduling and control are flexible to accommodate the many changes, such as arrivals of new orders, changes in due dates, and breakdowns. Bottlenecks occur frequently and often in different locations. The number of setups is high because of the many different products produced, but setup times are short.

Process Technology

General-purpose, flexible, sophisticated, computer controlled machines, robots, and material handling equipment are used in the FMS production system. Tooling, such as fixtures and pallets, are general purpose. Everything is automated and expensive. Because machines add most of the value to the product, process technology is very important. FMS production systems are usually technology leaders. An important part of the process engineering department's job is to remain current with new developments in process technology. Technological change is both incremental and revolutionary. The FMS production system is capital intensive. Material cost is high; direct labor is low. Indirect costs, especially those associated with support staff, are high.

Facilities

Facilities are medium in size, with limited economies of scale. Facilities are usually new and very clean. A great deal of infrastructure is needed in areas such as computer communication networks and maintenance and repair facilities. The speed of the process is medium. Because so many different products are produced, there are frequent, changing bottlenecks. Nevertheless, equipment utilization is relatively high, which is important because it permits the high capital cost of the FMS to be spread over as many parts as possible.

Manufacturing Outputs Provided by the FMS Production System

Cost and Quality

Three production systems can produce many products in low volumes—the job shop, batch flow, and FMS systems. Of these, the FMS provides the highest levels of the cost, quality, and delivery outputs. Quality is provided at a high level because expensive, sophisticated equipment is used and few if any operations are performed by operators. Even though the equipment and tooling are general purpose, they are so sophisticated that they can produce complex parts with tight specifications. A high level of manufacturing capability is required to operate an FMS production system properly because so much sophisticated technology is used. When the FMS is working properly, the equipment utilization is sufficiently high that products are produced at low cost. Because a small volume of each product is produced, only a modest opportunity for learning can occur. Consequently, the production system must be well managed to ensure that available learning opportunities are realized. An FMS is a line flow production system, so it provides the cost and quality outputs at line flow levels. However, these outputs are provided at the weakest levels of all the line flow production systems (see Figure 15-2) because the other line flow systems manufacture a smaller variety of products in much higher volumes using more specialized and less expensive equipment.

Performance

The FMS production system provides the performance output at the same high level as the batch flow and just-in-time production systems. This is a lower level than that provided by the operator-paced line flow system, but a higher level than that provided by the job shop and continuous flow systems. There are two reasons for this high level of performance: 1) The volumes of each product are such that the organization can commit a small amount of engineering resources to work with customers to design new, advanced features into the products; and 2) the equipment is so flexible and sophisticated that only a few changes, such as writing new computer programs and building new fixtures, need to be made to produce products with new features.

Delivery

Because many different products are produced at the same time, products will occasionally compete for time on the same equipment. This increases the time required to produce a product. Scheduling is difficult because numerous orders are in the plant at any one time. Therefore, it is relatively difficult to provide delivery times and delivery time reliabilities that are as good as those provided by other line flow production systems.

Flexibility and Innovativeness

Expensive, sophisticated equipment enables the FMS to provide a high level of flexibility and a slightly lower level of innovativeness. The FMS is designed to produce many different products in low volumes, so it is relatively easy to change products and volumes. It is a little more difficult to introduce new products and make design changes to existing products because new pallets, fixtures, and computer programs are needed. However, the FMS provides a higher level of innovativeness than the traditional line flow production systems with their specialized equipment, tooling, and relatively unskilled operators.

Machining and Assembly FMSs

There are two kinds of FMSs—machining and assembly. Machining FMSs are described in Situations 15.1, 15.2, 15.4, and 15.5. Assembly FMSs are described in Situations 15.3 and 15.6. These examples have three things in common: 1) The existing batch flow production system is replaced with an FMS because the batch flow system cannot provide the required levels of the market qualifying and order winning outputs; 2) quality is the order winning output, and cost and flexibility are the market qualifying outputs, and 3) the organizations wanted to gain technical and managerial expertise with the technologies in the FMS production system.

SITUATION 15.2

Aircraft Products Manufacturer Installs FMS[1]

IN 1984, LTV Aircraft Products installed an FMS to machine 1,300 different parts for the Rockwell B1-B bomber. The FMS was designed and built by Cincinnati Milacron. By 1988, the system had paid for itself in inventory savings alone. The flexibility of the FMS allowed the company to produce each of the 1,300 parts as needed and often in a batch size of one. In addition to reductions in inventory cost, labor costs decreased and product quality increased. It was anticipated that when the contract for the bombers ended, the FMS would be used to make parts for other aircraft as well as products for other industries.

SITUATION 15.3

An Assembly FMS at Perkins Diesel Engine Plant

PROBLEM[2]

In the early 1980s, the Perkins diesel engine plant installed over $120 million worth of new hard technologies, including an FMS for assembling cylinder heads. The FMS, installed as a turnkey operation by a U.K. automation house, assembled approximately 50 different cylinder heads for three-, four-, and six-cylinder engines in volumes averaging 13 units of each product per day.

The FMS consisted of eight robots, test equipment, parts feeders, a conveyor system, a wash station, and five separate computers. The robots inserted valves, valve springs, retaining caps, cotters, and oil seals, while the test equipment checked completed cylinder heads for leaky valves. Quality was checked automatically after each critical operation. The benefits from the FMS included the following: The workforce in the cylinder head assembly department was reduced from 30 to 18 employees; product quality improved, and delivery time was reduced; products were produced in lot sizes as small as one unit; and the company gained valuable experience with the new technologies in the FMS production system.

Analysis

The Perkins FMS is analyzed in Figure 15-3. Many products are manufactured in low volumes—50 products in volumes averaging 13 units of each product per day. Quality is the order winning output, and cost and flexibility are market qualifying outputs. Quality is very important for this type of product. Cost is also important, and a high level of flexibility must be provided because so many different products are produced. The FMS production system can best provide the market qualifying and order winning outputs for this mix and volume of products. Perkins has an adult level of manufacturing capability in facilities and process technology (because of the large investments it made in new technology) and an industry average level of capability in the other levers. Prior to implementation of the FMS, the company used a batch flow production system. Some of the adjustments that must be made to the manufacturing levers are explained below.

Human Resources

Training is required for production operators and personnel in maintenance, material handling, and other support departments.

Organization Structure and Controls

Maintenance procedures and activities in other support departments must be adjusted to support the FMS, so that all the potential benefits are realized.

Production Planning and Control

The current planning and control system was designed for a batch flow production system. It must be changed to take advantage of the FMS's flexibility.

Sourcing

Because products are produced more frequently and in smaller lots, delivery schedules for purchased parts will

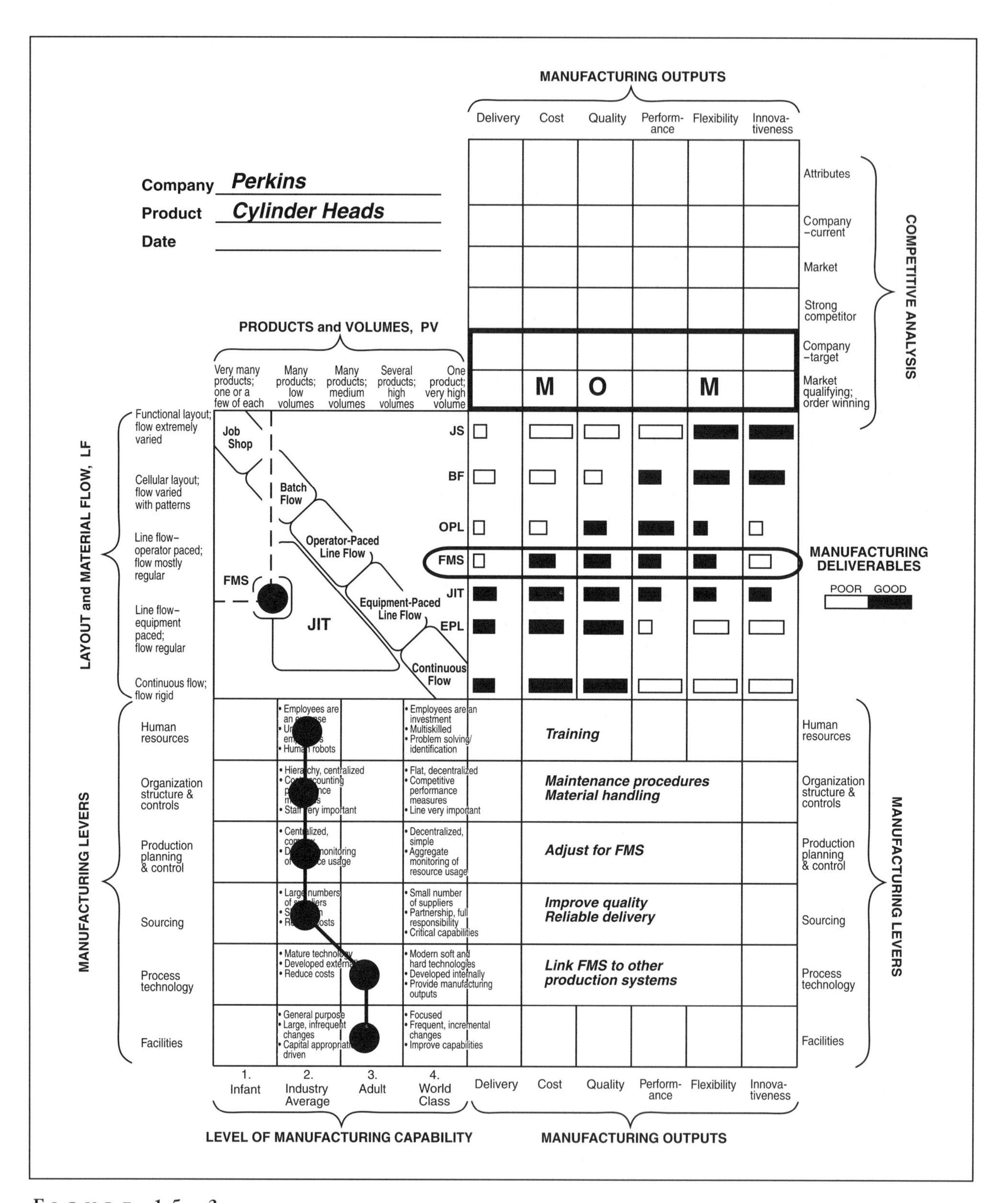

FIGURE 15–3

The FMS Production System at Perkins

change. The tooling and equipment in an FMS are so precise that they require high-quality materials and parts to work properly. Consequently, the quality of materials and parts received from suppliers must be improved.

SITUATION 15.4

A Machining FMS at Caterpillar[3]

IN 1989, Caterpillar, with sales of $11 billion, and Komatsu, with sales of $6 billion, were engaged in a worldwide struggle for the heavy equipment market. Caterpillar never lost its reputation for quality or service during the struggle. Its costs were too high, however, and it knew that cost cutting alone would not make it competitive. So the company decided to redesign its factories at a cost of more than $2 billion. Caterpillar nicknamed the plan Plant-with-a-Future: PWAF.

One plant, the KK plant, produced 120 types of transmissions for the whole catalog of Caterpillar machines. In the KK plant's machining department, a long line of 35 machine tools, each requiring its own operator, stretched down the left side of the main aisle. On the right side of the aisle 4 (of an eventual 32) flexible machining systems were installed. The machining systems were part of an FMS production system. The machines on the left side of the aisle were part of Caterpillar's old batch flow production system. Because the machines in the batch flow system were older and could produce only one kind of transmission case at a time, the area was crowded with bins of cases waiting to be machined. When a batch was finished, the operator spent anywhere from four hours to two days setting up for the next batch. Once done, he or she might need two or three tries before the tool ran correctly. With adjustments such as these made at 35 stations, many $1,000 cases ended up in the scrap heap.

The four flexible machining systems in the FMS production system did the same work of milling, drilling, boring, tapping, deburring, and reaming. However, they were programmed to handle any case the plant made, and the setup time was only a few seconds. The system selected the right tools from a rotating belt and inserted them into spindles. And because the tool did the job right the first time, there was little scrap. Tool wear was checked electronically by monitoring the torque on the tool spindle. Before a cutting tool broke or cut off size, the spindle stopped and the worn tool was replaced automatically.

The other departments in the KK plant—gearmaking, heat treatment, assembly, final testing—were all converted to the FMS production system. By 1990, the time to build a transmission had been reduced from three months to 15 days, and costs had decreased significantly. As Figure 15-4 shows, the KK plant changed its production system from a batch flow system to an FMS system because it needed to provide products at lower cost, with better delivery, while maintaining or improving quality. Not only did Caterpillar have to change its production system from batch flow to FMS, it also had to create an FMS production system with above average manufacturing capability. The competition it faced from Komatsu was such that the required levels of the market qualifying and order winning outputs could be provided only by a production system with an adult or higher level of manufacturing capability.

SITUATION 15.5

An FMS Production System at Pratt and Whitney

IN 1980, Pratt and Whitney opened an FMS plant in Halifax, Nova Scotia. Seventy different products in volumes ranging from 30 to 1,000 units per year were machined in the plant. Pratt and Whitney decided to use an FMS production system because it required the quality and cost of a line flow system but, with the high number of products and the low volumes, the company could not use an operator- or equipment-paced line flow or JIT system. The company also wanted to gain valuable experience with the new technologies in the FMS production system.

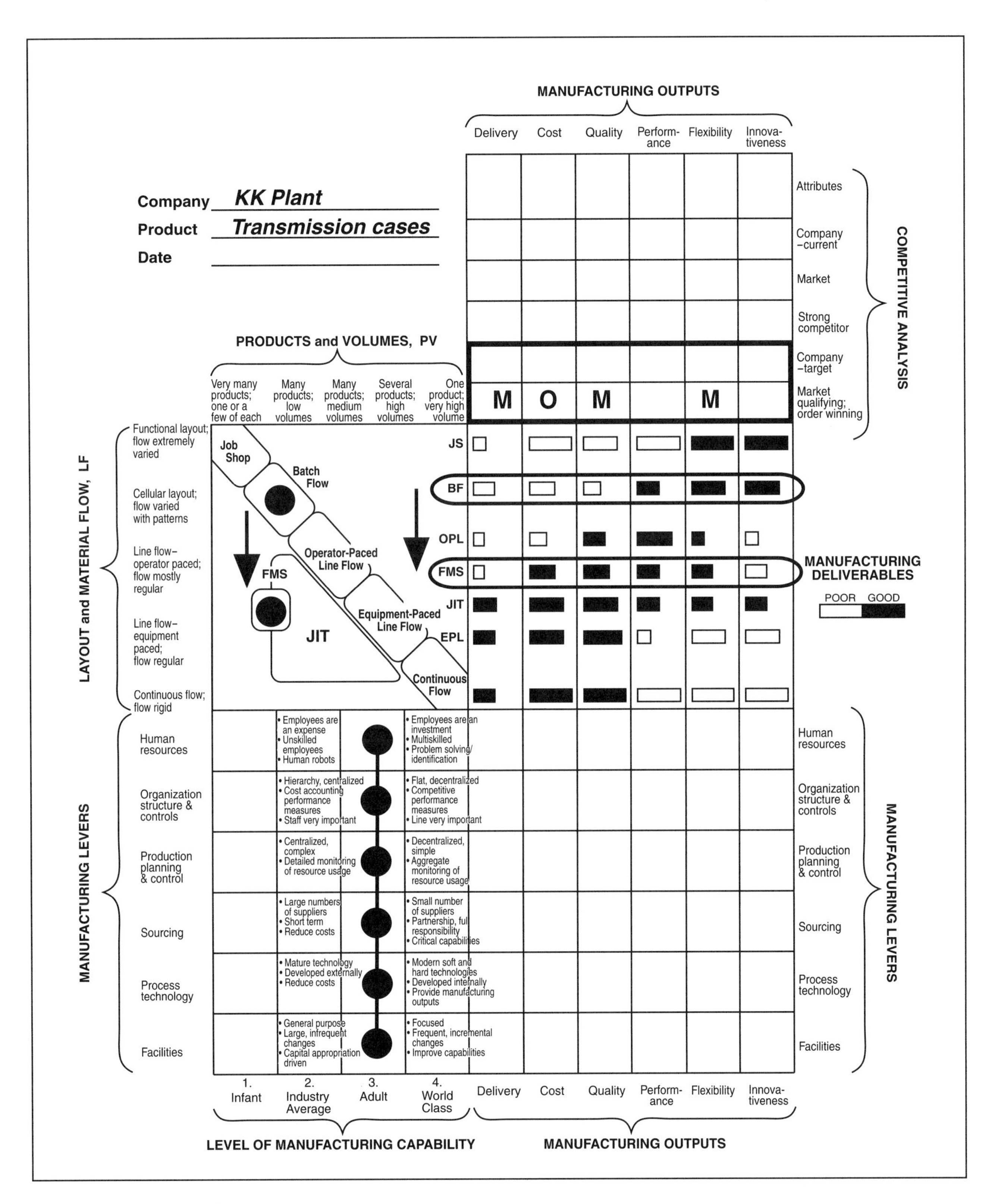

FIGURE 15–4

The FMS Production System at the Caterpillar KK Plant

Situation 15.6

Allen-Bradley Showcases Its FMS[4]

ALLEN-BRADLEY manufactures automation controls—everything from contactors and relays for starting motors to programmable logic controllers for directing the activities of flexible manufacturing systems. The company found that the production system it used to manufacture motor contactors and control relays was increasingly unable to provide the levels of cost and quality required to be competitive in the world market. At the same time, one part of Allen-Bradley's corporate strategy sought to increase the company's sales of components used in factory automation. Allen-Bradley had difficulty achieving this part of the strategy because it could not convince many customers to adopt its ideas for flexible manufacturing systems and factory automation.

In the early 1980s the company decided to build the kind of FMS it had been trying to sell its customers and use that production system to make motor contactors and control relays. A few years later, its first FMS production system was ready. It produced 125 variations of contactors and relays. Ten years and many improvements later, the FMS production system could manufacture 937 variations at a rate of 600 units per hour, in batches as small as one, and with no direct labor. Changeover was automatic and took less than six seconds, which was the cycle time of the machines in the FMS. The company believed that its FMS production system was the lowest-cost producer of contactors and relays in the world. Lower cost, improved quality, and the ability to respond quickly to customer needs were helping Allen-Bradley take orders away from its competitors.

Notes

1. Adapted from Brandt, R., and O. Port, "LTV Aircraft Products Group: Only the Beginning," *Industry Week*, pp. 50–54, March 21, 1988.

2. Adapted from Wylie, P., "Perkins £50m Drive Aims for Diesel Lead," *The Engineer*, November 29, 1984, p. 36.

3. Adapted from *Fortune*, pp. 59–60, May 21, 1990.

4. Adapted from "Case Study: Allen-Bradley," *IEEE Spectrum*, pp. 37–39, September 1993.

CHAPTER 16

The Operator-Paced Line Flow Production System

Product Mix and Volumes

The operator-paced line flow production system produces many similar products in medium volumes on equipment arranged in a line layout (see Figure 16-1). The volume of each product is higher and more regular than it is in the batch flow production system, but is lower and more variable than it is in the equipment-paced line flow system. Compared to the just-in-time system, the operator-paced line flow system produces fewer products in higher volumes. However, the variability in the product volumes is about the same.

Layout and Material Flow

A family of products with somewhat similar manufacturing requirements is produced on equipment arranged along a line. Many organizations use line layouts, and it is useful to differentiate between lines where the pace or speed of the line is set by the operators and lines where the pace is set by the equipment. In operator-paced line flows, the rate of production depends on the number of operators assigned to the workstations on the line, the speed at which the operators work, and how well they work together as a team.

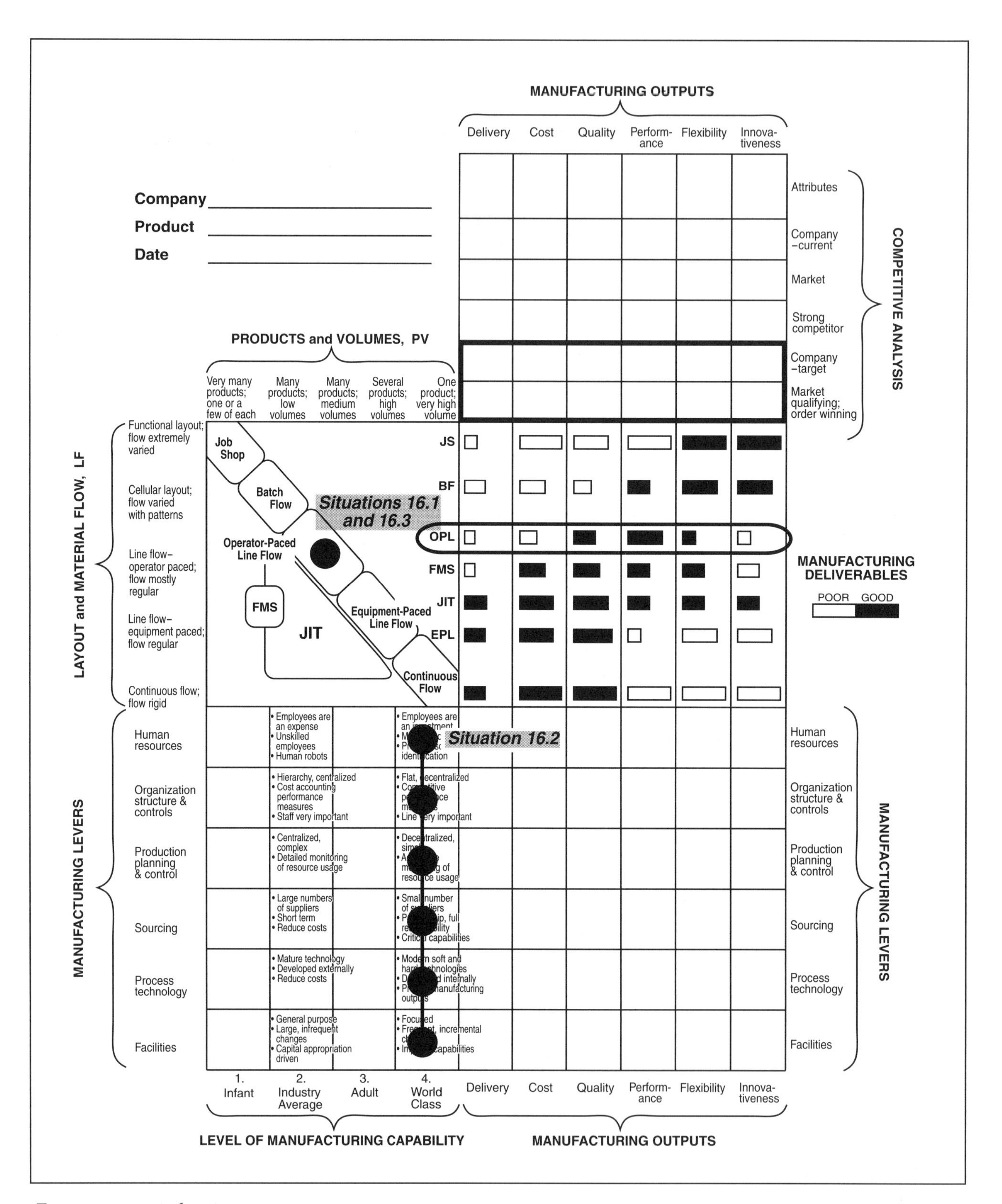

FIGURE 16–1

The Operator-Paced Line Flow Production System

The material flow in an operator-paced line flow production system is regular for the most part. There will be some variations from product to product, and from time to time, as requirements for products change. Equipment and tooling are assigned to the family of products being produced on the line, so they are more specialized than in the job shop and batch flow production systems. However, the product family is large and the amount of specialization is less than is required in the equipment-paced line flow system.

Competitive Advantage

The operator-paced line flow production system provides good levels of all the manufacturing outputs. It is used when the number of different products is too high and the production volume is too low or too variable for an equipment-paced line flow production system. The line is more flexible than an equipment-paced line and can be run at different speeds. One of the best-known operator-paced line flow production systems is the one used at McDonald's restaurants. A limited variety of different products are produced in volumes that vary during the day. Many products are produced on the same line, and line speed can be changed quickly by adjusting the number of operators on the line. Teamwork is stressed.

Manufacturing Levers in the Operator-Paced Line Flow Production System

Human Resources

Production operators are less skilled than those in the job shop, batch flow, or just-in-time production systems, and more skilled than those in the FMS, equipment-paced line flow, or continuous flow production systems. Procedures and standards for performing all production operations are carefully developed, improved, and updated. They are necessary for preserving quality and maintaining a steady production rate on the line. Operators are assigned to stations along the line where they perform similar operations on the different products produced on the line. Operators are trained to work at more than one station, and sometimes they do routine maintenance on their equipment. They also participate in problem solving and improvement activities. Incentive wage schemes can be used. One element of these schemes is a reward that depends on the performance of the entire team. Many small staff groups provide services such as engineering, production planning and control, maintenance, and quality. The groups are small and quite responsive to the needs of the production line.

Organization Structure and Controls

Job shop, batch flow, and operator-paced line flow production systems are usually profit centers. As far as corporate influence is concerned, they are relatively autonomous. The organizational structure is not as flat as it is in the job shop and batch flow production systems, but it is not as hierarchical as it is in the FMS, equipment-paced line flow, and continuous flow production systems. The organizational structure is similar to that found in the just-in-time system. Two of the most important problems in the operator-paced line flow production system are 1) managing materials, and 2) scheduling production so that the line is used efficiently, the many different products are produced on time, setups and inventories are minimized, and other management objectives are achieved.

Sourcing

An operator-paced line flow production system has a moderate amount of influence over its suppliers for many reasons. Many different products are produced in medium volumes. Customer demand and hence production are irregular. Product life cycles are short, which means that products are redesigned frequently and there is a rapid rate of new product introductions. All this results in small and irregular orders to suppliers, making it difficult to influence or control them. Nevertheless, operator-paced line flow production systems try to work with suppliers to improve cost, quality, and delivery. In particular, high quality and fast, reliable deliveries are necessary to keep the line from stopping. Influence over suppliers is higher than it is for the job shop and batch flow systems, but it is not nearly as high as it is for the equipment-paced and continuous flow production systems.

Production Planning and Control

Production is make-to-order. Sometimes products are produced early and placed in a finished goods inventory to smooth workloads in the plant. Raw materials inventory is large to ensure that the line will not stop because of a shortage of purchased material. Work-in-process inventory is low because products are produced quickly on the production line. Finished goods inventory is also low. An operator-paced line flow production system is very flexible. Consequently, scheduling is flexible and frozen schedules are short, if they exist at all. Little expediting is needed, except when shortages of purchased materials threaten to stop the line.

Process Technology

Equipment, fixtures, and tooling are not general purpose like they are in the job shop, batch flow, and FMS production systems, nor are they as specialized as they are in the equipment-paced line flow and continuous flow production systems. Production runs are long for some products and short for others. Because many products are produced on the same line, short changeover times are important. Reducing changeover times and scheduling to minimize time lost due to changeovers are important management concerns. Cooperation between process engineers and product design engineers is necessary to ensure that new products, new product features, and design changes to existing products can be introduced effectively. Operator-paced line flow systems are usually technology followers. They are not as common as batch flow and equipment-paced systems, so they follow technology developments in these, modifying them as needed.

Facilities

Facilities are medium in size with limited economies of scale. The speed of the process is medium. The process is neither labor nor capital intensive. Equipment along the line is relatively balanced. Consequently, there are usually no bottlenecks, and equipment utilization and labor efficiency are high.

Manufacturing Outputs Provided by the Operator-Paced Line Flow Production System

Cost and Quality

Because equipment and tooling are somewhat specialized for the products produced on the line, and the operators are well trained to perform the small number of tasks required of them, the cost and quality of products are better than they are in the job shop and batch flow production systems. In addition, products are produced in medium volumes on a somewhat regular basis, which enables learning to occur and improvements to be made. This also contributes to better levels of cost and quality.

Performance

The operator-paced line flow production system provides the highest possible level of performance for three reasons:

1. The volume of each product and product family is sufficiently high to justify the research and development needed to design a steady stream of new features.
2. The volumes are high enough to support the engineering work needed to design new production processes to produce the new product and improve existing processes to produce existing features more efficiently.
3. The combination of operators, equipment, and tooling that are somewhat specialized but still flexible means it is relatively easy to make whatever changes are needed for the stream of new features and new processes.

Delivery

Delivery is provided at a high level because all the operations required to produce a product are done on equipment arranged along a line. Once production of a product starts, little time passes before it is finished.

Flexibility and Innovativeness

The levels of flexibility and innovativeness provided by the operator-paced line flow production system are not as high as those provided by the job shop and batch flow production systems. A smaller number of different products are produced on equipment and tooling that is relatively more specialized. This makes it a little more difficult to change products and volumes, introduce new products, and make design changes to existing products. However, the operators, equipment, and tooling are much less specialized than they are in the equipment-paced and continuous flow production systems. Consequently, the operator-paced line flow production system provides higher levels of flexibility and innovativeness than these systems.

SITUATION 16.1

Sun Microsystems Builds New Operator-Paced Line Flow Production System[1]

SUN MICROSYSTEMS builds computer workstations for many different applications. The company was founded in 1982 and went from $8.6 million in sales in 1983 to $527 million in 1987. However, manufacturing was holding back sales because it could not meet delivery dates. The problem was an outdated plant and production system. So Sun decided to build a new plant. The new assembly line would have the flexibility to build three major product families, each of which had many options, in any order. For example, a small desktop computer with 4 megabytes of memory could be followed by a large desk-side computer with 2.3 gigabytes of memory, which could be followed by a cabinet-size network server. To make the line work, Sun needed special roller conveyors that delivered kitted subassemblies to the line; operators who were multiskilled to perform many operations; and a computerized scheduling and control system that tracked operator and material availability, the status of each order, etc.

Once the new operator-paced line flow production system was implemented and working, Sun experienced a 150 percent increase in capacity and tripling of its net income. Delivery reliability increased dramatically and Sun had the flexibility to respond to changing customer needs.

SITUATION 16.2

Lincoln Electric's Operator-Paced Line Flow Production System[2]

THE LINCOLN ELECTRIC COMPANY of Cleveland is an excellent example of a well-managed operator-paced line flow production system. Lincoln Electric was founded by John C. Lincoln in 1895 as a repair shop for electric motors. Before long, the company was manufacturing its own brand of electric motors. By

1912, it became interested in welding equipment when welding began to gain widespread acceptance as a process for joining metal. From these beginnings, Lincoln Electric became one of the world's largest manufacturers of welding machines and electrodes. Lincoln Electric used an operator-paced line flow production system from the beginning. The improvements made to the system over the decades raised the level of manufacturing capability of the production system to the world class level. It is interesting to note some of the milestones in the development of Lincoln's operator-paced line flow production system.

In 1914, James F. Lincoln took over leadership of the company. One of his early actions was to ask employees to elect representatives to a committee that would advise him on company operations. During the advisory board's first year, working hours were reduced from 55 to 50 hours per week. In 1915, each employee was given a paid-up life insurance policy. In 1918, an employee bonus plan was attempted but did not catch on. It succeeded when it was tried again in 1934. The first annual bonus amounted to about 25 percent of wages, and there has been a bonus every year since. A suggestion plan was started in 1929. By 1944, Lincoln employees enjoyed a pension plan, a policy of promotion from within, and continuous employment. Basic wage levels for jobs at Lincoln were determined by a wage survey of similar jobs in the Cleveland area. They were adjusted quarterly for changes in the Cleveland area consumer price index. When possible, basic wage levels were translated into piece rates. Practically all production workers and many other employees were paid by piece rate.

Welding machines and electrodes are manufactured at Lincoln's main plant (see Figure 16-2). The operator-paced line flow production system that manufactures the welding machines works as follows. Materials flow from a dock on the north side of the plant through the production lines to a small storage and loading area on the south side of the plant. Materials used at each workstation are stored as close as possible to the workstation. The lines are designed so that several different welding machines can be produced on the same line by adjusting the line speed, changing the assignment of operators to workstations, and using different tooling. Piece rates are paid to teams of operators rather than to individuals so that operators work together as a team.

FIGURE 16–2

Plant Layout at Lincoln Electric

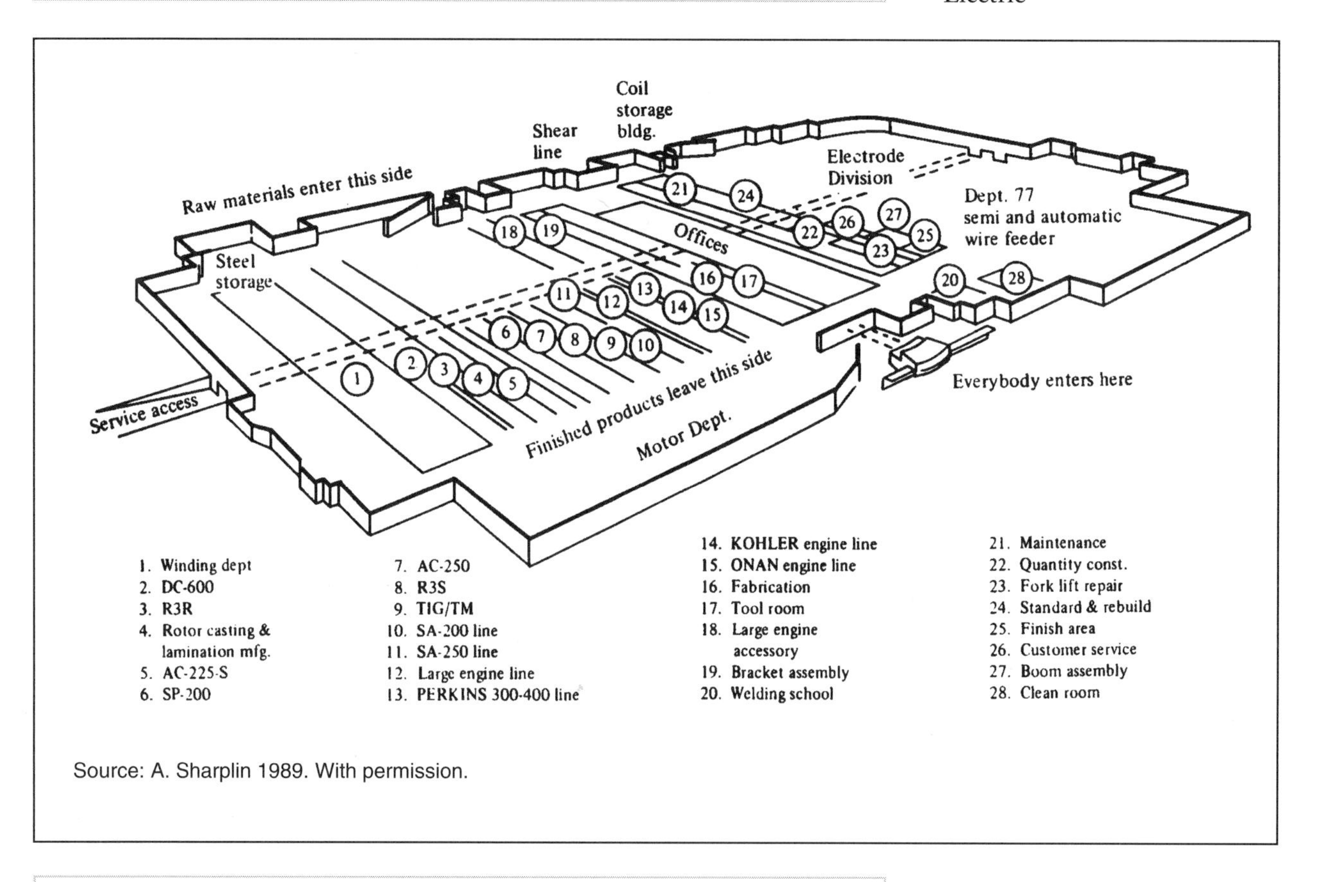

Source: A. Sharplin 1989. With permission.

SITUATION 16.3

European Manufacturer Improves Operator-Paced Line Flow Production System

NDL COMPANY, a European manufacturer of transportation equipment costing over $80,000 per unit, wondered whether it should change the way it competed against its two large competitors. The competitors, who dominated the market, offered customers standard products with a few options. Quality was very high, and prices were quite reasonable. NDL, on the other hand, produced many models and options. There were so many combinations of models and options that production managers joked about how the company had never produced the same product twice. Occasionally, a new combination of options created problems on the assembly line when it was discovered that the particular combination could not be produced because of interferences between parts. Customers were fiercely loyal to NDL. They knew they could get precisely what they needed for their unique applications. NDL used an operator-paced line flow production system.

Human Resources at NDL

The variety of models and options in the products on the production lines caused the workloads in the stations to vary. NDL responded by employing operators who were trained to do many tasks. The speed of the lines was increased or decreased to meet production requirements by adding or removing operators. Production lines were divided into sections, and a team of operators was assigned to each section. Each team was responsible for completing all the operations at the stations in their section. Each team had a leader who assigned operators to stations. Depending on what products were scheduled to be produced and the skills of the team members, operations could be shifted to different stations or shared between stations. More than one operator was sometimes assigned to the same station. Operators at adjacent stations often helped each other.

In addition to an hourly wage rate, the company also had an incentive plan and profit-sharing plan. With the incentive plan, each member of the team received an equal share of a set price for each product produced without defects. The team was penalized a set cost, equal to about approximately ten times the set price, for each defect it produced. The set cost pushed employees to stop the line rather than produce a defect. The incentive plan also awarded each employee an attendance bonus, which worked as follows. The employee received *x* cents an hour if he or she came to work on time, another *y* cents an hour if he or she came to work on time every day for one week, and another *z* cents an hour if he or she came to work every day for four consecutive weeks. Under the profit-sharing plan, about 30 percent of the pretax, prebonus profit became a bonus pool. Each employee's share of the bonus pool was determined by a semiannual merit rating that measured the employee's performance. Four factors figured in the merit rating—dependability, quality, output, and ideas and cooperation.

There were few job classifications. All production operators received the same hourly wage, regardless of what tasks they were assigned. Employees moved back and forth between departments with a minimum of red tape.

Organization Structure and Controls at NDL

NDL had a very flat organizational structure. Relatively large staff groups were responsible for product design, process engineering, materials, maintenance, and so on. Much of the staff groups' time was spent managing the many models and options produced. Every staff group employee and all senior managers spent one day a year working on the production lines, and another day working with customers in the sales department. No one exhibited an "it's not my job" mentality anywhere in the organization. Effective communication was stressed. Teams met three times a week to assign work and discuss problems.

Production Planning and Control at NDL

Production of final products was make-to-order. There was no finished goods inventory. Many common parts were make-to-stock; that is, they were built to a forecast and inventoried. Large inventories of raw materials and purchased components were kept. A modern MRP II computer system was used to plan production, control materials, and control and monitor activities in the plant.

Sourcing at NDL

NDL purchased many different materials and parts in medium volumes from its large supplier base. NDL had less influence over its suppliers than it would have liked and found it difficult to reduce the cost of purchased materials and parts, and improve the quality and delivery provided by suppliers. To increase its influence over its suppliers, NDL attempted to reduce the supplier base. It eliminated poor suppliers and gave more business and longer contracts to those suppliers who had high levels of manufacturing capability and a record of providing high levels of the manufacturing outputs. NDL wanted to work with suppliers to reduce cost, and improve quality and delivery of all purchased materials and parts. NDL also wanted suppliers with skills in product design so that they could help the company provide high levels of performance and innovativeness.

Process Technology at NDL

The equipment, tooling, and fixtures used in the production process were neither general purpose nor specialized. Some of the equipment and most of the fixtures were designed and built by NDL to be flexible enough for all the different models and options produced. Like all line flow production systems, there was a great deal of automation at the workstations. However, almost every workstation required operators to perform many tasks, several of which varied from model to model and option to option. Communication between product design engineers and process engineers was good. Both groups were located at the same site, albeit in different buildings. A formal design for manufacturability program was in place. Process engineers were involved early in the design process, which reduced the time needed to introduce new options and models. It also reduced production costs and improved quality because parts were designed so that they were easy to manufacture using the existing production processes.

Facilities at NDL

The NDL plant was large, modern, and clean. The line speeds were slower than in other plants where standard products were produced on equipment-paced line flow production systems. Fabricated parts were produced in the other departments using batch flow and job shop production systems.

Notes

1. Adapted from M. A. Vonderembse and G. P. White, *Operations Management*, St. Paul, MN: West Publishing Company, p. 267, 1991.
2. Adapted from Arthur Sharplin, "Lincoln Electric: 1989." Used with permission from Arthur Sharplin, Austin, TX.

CHAPTER 17

THE JUST-IN-TIME PRODUCTION SYSTEM

Just-in-time (JIT) is a production approach wherein producing products and making improvements to the production system are equally important objectives. Improvements are made by identifying and removing wastes, for the purpose of reducing cost and improving quality, performance, delivery, flexibility, and innovativeness. JIT is used by many manufacturers. For some, JIT is a collection of techniques for improving a production system. For others, it is an entirely new production system.

A Collection of Techniques

Since the mid-1980s, new techniques have been developed and old techniques have been rediscovered for improving production systems. These include the improvement approaches in Chapter 9, and the soft and hard technologies in Chapter 10. Often called "JIT techniques," they can be implemented in most production systems.

A New Production System

Over a 20-year period beginning in the early 1950s, the Toyota Motor Company of Japan developed a new production system for producing automobiles, with great success. The Toyota production system produced cars with higher quality, at less cost, and in less time than the traditional line flow and batch flow production systems used by Toyota's competitors. When an oil embargo plunged Japan into a deep recession in the early 1970s, many Japanese companies, desperate to cut costs, started to implement the new Toyota production system. The results were startling—and the rest is history.

The Toyota production system has continued to evolve and spread. It is now used in industries far removed from the automotive industry. Now, it goes by several different names: just-in-time, lean production, stockless production, zero inventories, when it's needed, and so on. Although the new production system incorporates all the JIT techniques, it is considerably more than a collection of techniques. It is a production system that by its nature enforces continual improvement of the production system by identifying and eliminating wastes.

Many companies are incorporating JIT techniques into their production systems. A few are changing their production systems to JIT production systems—usually when they have no choice—because JIT is the only production system capable of providing the cost, quality, performance, delivery, flexibility, and innovativeness outputs at the levels required by their customers.

Products and Volumes

A JIT production system is a line flow production system that produces many products in low to medium volumes (see Figure 17-1). Before the development of the JIT production system, this product mix was produced on a batch flow system. The traditional line flow systems could not be used because the number of different products was too high and volumes of any particular product were too low.

Layout and Material Flow

The JIT production system was developed so that this mix and volume could be produced on a line flow production system. The objective was the lower cost of line flow production compared to batch flow production. Toyota started to develop JIT in 1950. At that time, Toyota produced small volumes of many models of cars—each designed to meet the requirements of different domestic and foreign markets. Toyota found that its batch flow production system could not provide the levels of cost

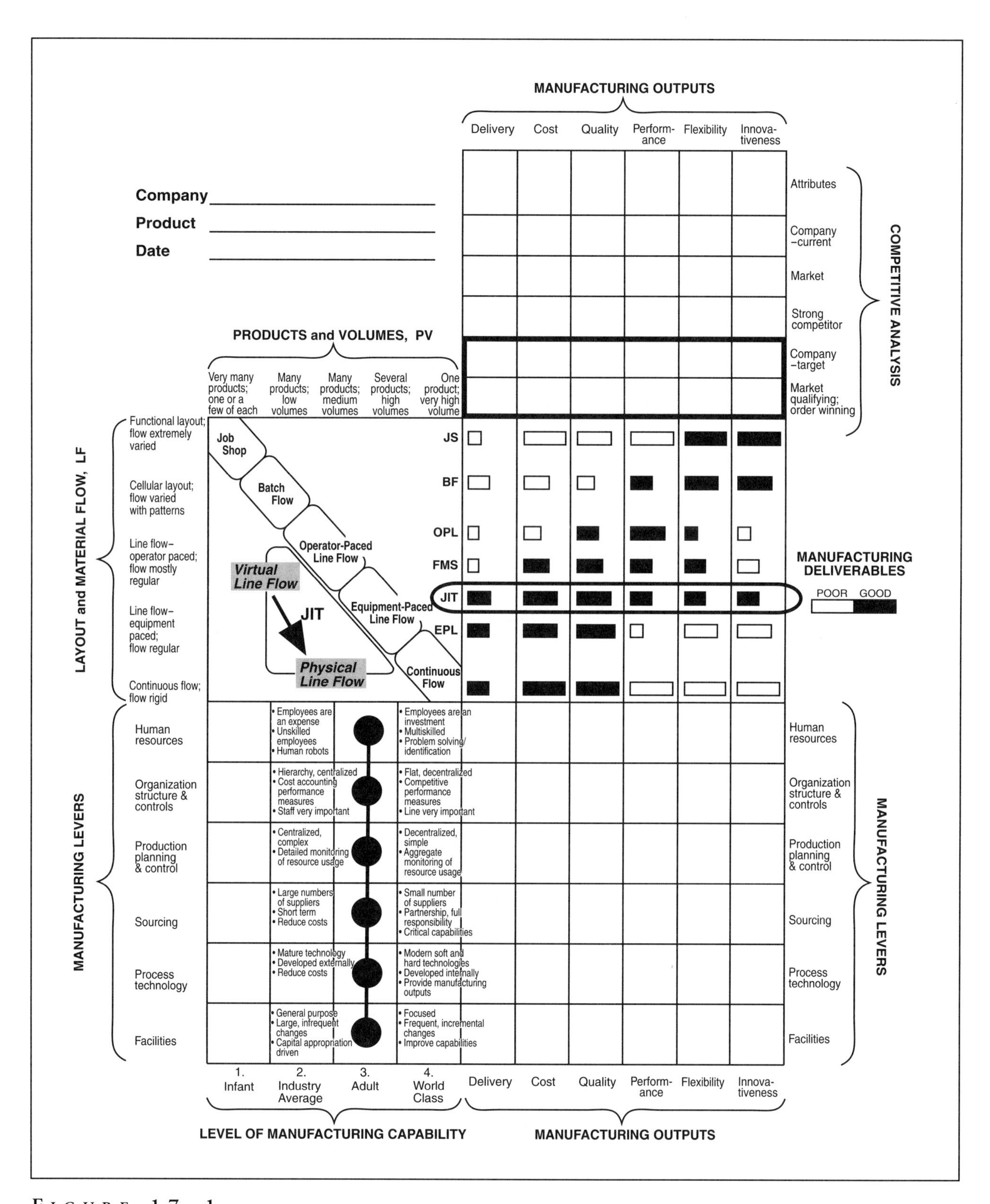

FIGURE 17–1

The Just-In-Time Production System

and quality provided by the line flow production systems of its larger, foreign competitors. Toyota could not use a traditional line flow production system because it was producing too many different products in small volumes. The company's response was to develop a new line flow production system that would produce many products in low to medium volumes. Taiichi Ohno, the former vice-president of Toyota and father of just-in-time, gave some of the credit to Henry Ford for the work Ford did in the 1920s at his River Rouge plant in Detroit. Said Ohno, "I think that if the American king of cars were still alive, he would be headed in the same direction as Toyota."[1]

As Figure 17-1 shows, the JIT production system occupies a large region on the PV–LF matrix. Different points in this region represent different degrees of implementation of JIT. As we will see, the top part of the region uses a "virtual" line flow, while the bottom part of the region is a "physical" line flow. Implementing JIT is a journey, consisting of many small steps that form a path through the JIT region. The path starts in the upper left corner and ends in the lower right.

Competitive Advantage

The JIT production system is a line flow production system that produces many products in low to medium volumes. By its nature, the JIT system forces wastes to be eliminated. It provides the cost, quality, and delivery outputs at the levels associated with traditional line flow production systems, while still providing flexibility and innovativeness at levels associated with the batch flow production system. The JIT production system uses the following mechanism to identify and eliminate waste:

1. All inventory is moved from the stockroom to the plant floor and is located in those areas where the parts in the inventory are used and where they are produced. Maximum levels are set for the amount of inventory of each part that can be held at each location.
2. Management lowers the maximum inventory levels at those areas of the production process where it wants to make improvements. This strains the area as it tries to produce without the cushion of a large backup inventory.

3. Soon problems develop that prevent the work area from completing its work before the inventory runs out, and a shortage occurs. The shortage causes a production stoppage. A production stoppage is a call to action. Extra resources are mobilized to seek out the problems or wastes causing the stoppage. The problems or wastes are studied carefully. Root causes are identified. Permanent solutions are found and implemented so that these problems do not recur. Typical solutions include the following: Setup times are reduced to facilitate small lot production; greater use is made of quality control techniques to prevent defective products; work procedures are improved; tooling and materials are improved; scheduling is improved; employees are trained to do multiple tasks; and product design is improved so that products are easier to manufacture. *Note that as soon as a problem is identified, the inventory level is increased to its original level so that production can continue without further disruptions while personnel find and implement a permanent solution for the problem. When the solution is implemented, inventory is reduced again to expose the next problem.*
4. This routine of removing inventory to strain different areas of the production process to force them to identify and eliminate wastes is repeated again and again until all waste is eliminated from the process. The analogy of water flowing over rocks is used to describe this process (see Figure 17-2). The water represents inventory and the rocks represent problems. The rocks or problems are exposed when the water or inventory level is lowered.

The JIT production system is the most difficult of the seven production systems to design, implement, manage, and operate. The primary reason is that, unlike other production systems that simply try to produce a particular mix and volume of products, JIT seeks to do two things: 1) produce many products in low to medium volumes with high levels of the manufacturing outputs, and 2) continuously identify and remove waste from the system. This requires very high levels of manufacturing capability in all

FIGURE 17–2

Reducing Inventory to Identify and Eliminate Waste

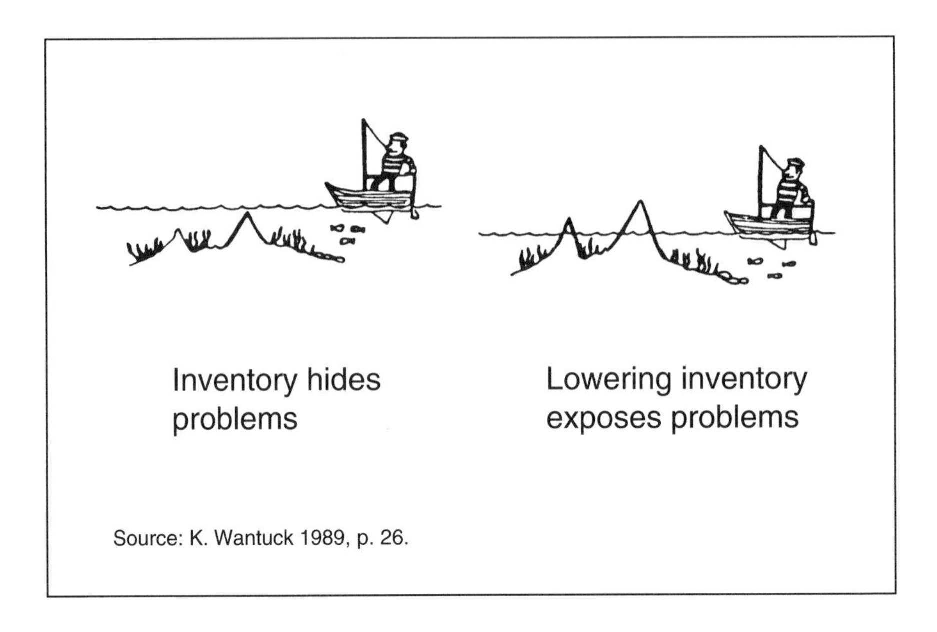

the manufacturing levers. Employees, suppliers, equipment, and systems must have very high capabilities. Toyota took more than 20 years to implement JIT, which is evidence enough to prove its difficulties. The conventional wisdom is to use the JIT production system when the following three conditions hold:

1. Many products are produced in low to medium volumes. The current production system is a batch flow system, and a traditional line flow production system cannot be used because too many products are produced, in volumes that are too low.
2. Higher levels of the cost, quality, and delivery outputs are required than what can be provided by the batch flow production system, even when it has a high level of manufacturing capability.
3. The level of manufacturing capability is high.

Today's products are offered in a wider array of options. Products are often tailored to meet the needs of small-niche markets, resulting in shorter product life cycles. Customers are also demanding better levels of cost, quality, and performance. The JIT production system is ideally suited for this environment because it can produce this variety of products in volumes that are too low for the traditional line flow production systems.

SITUATION 17.1

Davidson Instrument Panels Implements JIT[2]

DAVIDSON INSTRUMENT PANELS is located in New Hampshire, about 900 miles from its automotive assembly plant customers in Detroit. Davidson recently changed its batch flow production system to a physical line flow JIT production system (see Figure 17-3). The plant layout and material flow was changed from a functional layout with production in batches to product-focused lines. Setup times were reduced from 8 hours to 20 minutes. The production rate for each line could be synchronized with customers' requirements, permitting products to be produced and shipped just in time. Programs for statistical process control, voluntary improvement teams, returnable containers, stablization of schedules, and so on, were started.

Davidson was pleased with its JIT production system. Product costs decreased, quality improved, and manufacturing cycle times dropped. Inventory turns, for example, increased from 7.5 to 28 turns per year over a four-year period.

FIGURE 17–3

The JIT Production System at Davidson

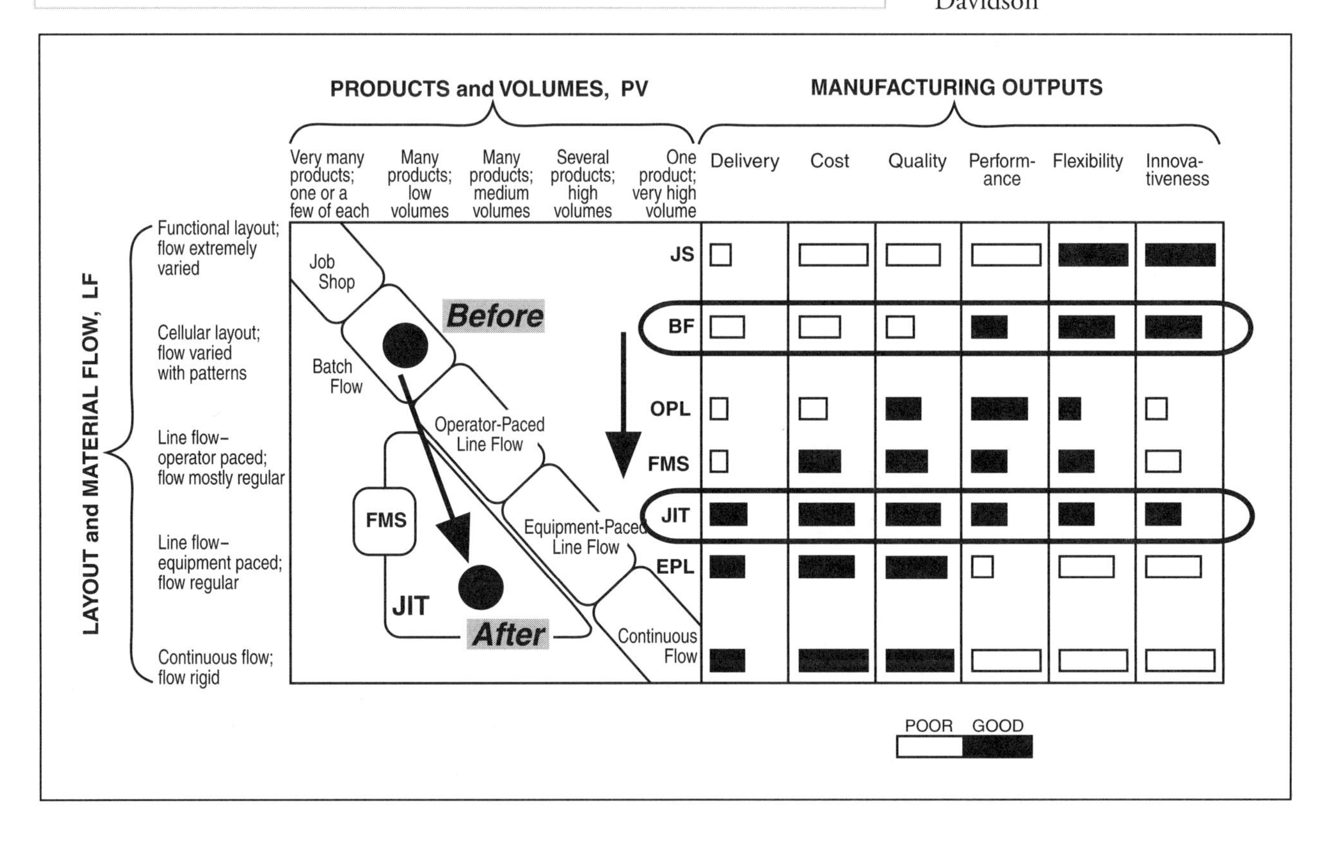

SITUATION 17.2

Heat Exchanger Manufacturer Is Disappointed with JIT

LNG COMPANY manufactured heat exchangers at a plant located near Toronto. On learning about JIT, LNG decided to relocate some production equipment from four areas of the plant into a physical line flow JIT production system to produce a family of heat exchangers (see Figures 17-4 and 17-5). Equipment was arranged into a U-line. Workstations were designed to eliminate unnecessary effort and allow work to be shared among stations. Operators were trained to do more than one job. Tools and fixtures were standardized. Buffer inventories between workstations were eliminated.

Problems developed as soon as the new line started. Inadequate procedures and quality problems precipitated by poor tooling and materials caused stoppages. There were material shortages. When equipment broke down, repairs were lengthy, and the entire line was stopped for long periods. All these problems resulted in late deliveries to customers and increased costs. After eight months of trying to make JIT work, the JIT project was put on hold. Large buffer inventories were created to decouple workstations so that production activities could continue when breakdowns occurred or scrap was produced. Batch sizes were increased to previous levels, and the company reverted to its old batch flow production system.

There are many reasons why companies like Davidson can move from a batch flow production system to a physical line flow JIT production system, and companies like LNG cannot. Davidson had a higher level of manufacturing capability when it undertook changing its production system. Davidson had a more effective implementation plan than LNG, and its execution was better (see Chapter 7). It took Davidson four years to achieve the desired results, while LNG stopped its JIT implementation after less than one year. When problems were exposed at Davidson, labor and other resources were available to find and implement

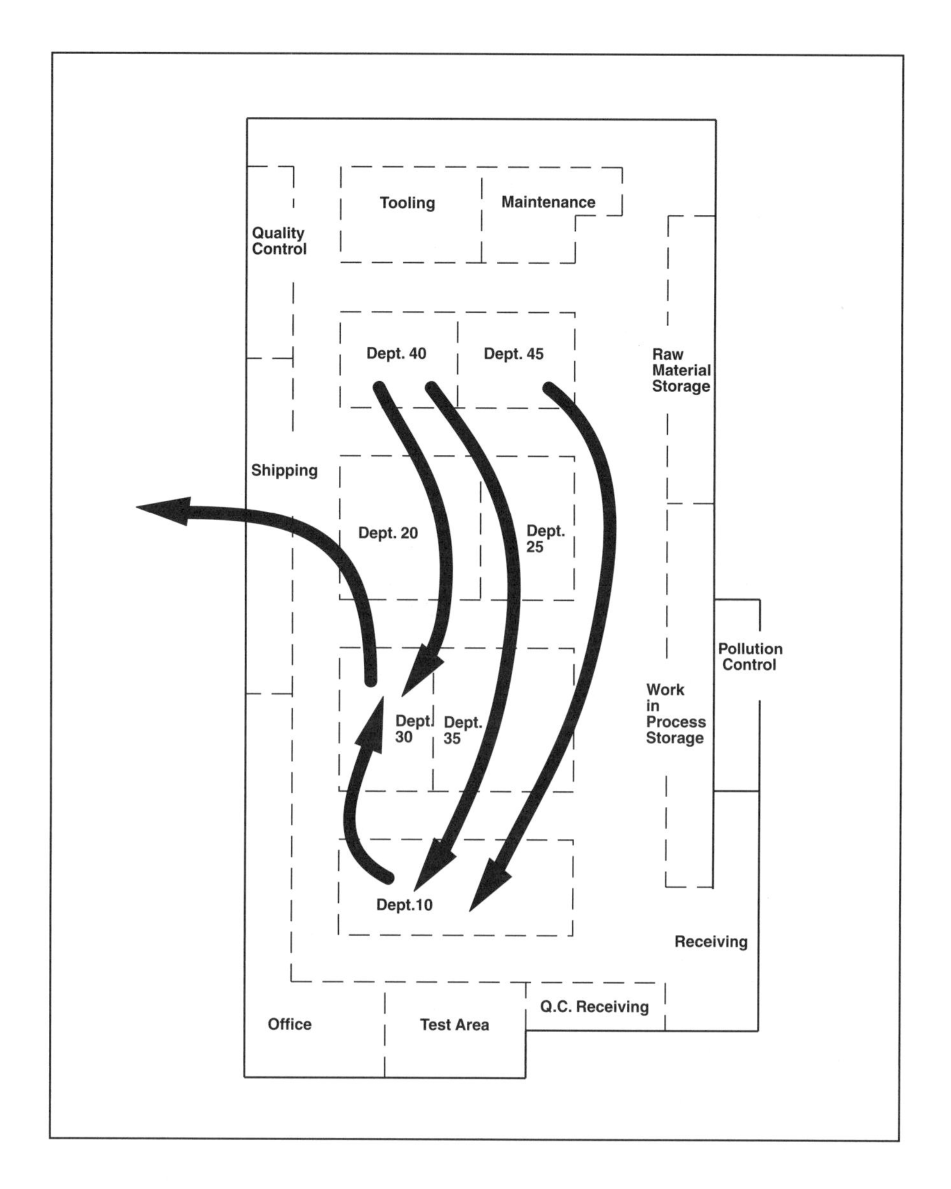

FIGURE 17–4

Layout and Material Flow at LNG for Heat Exchanger Family before JIT

solutions. These resources were not available at LNG. Problems were exposed but were not solved, and so the JIT production system could not work.

Regardless of the level of manufacturing capability, the thoroughness of the implementation plan, and the determination to stay the course, jumping from a batch flow production system to a physical line flow JIT production system is usually not a good idea. The number of adjustments to the manufacturing levers and the size of the adjustments are such that, unless the organization has a very high level of manufacturing capability, the chances of failure are high. The conventional wisdom is to implement JIT gradually by moving from the batch flow production system to a

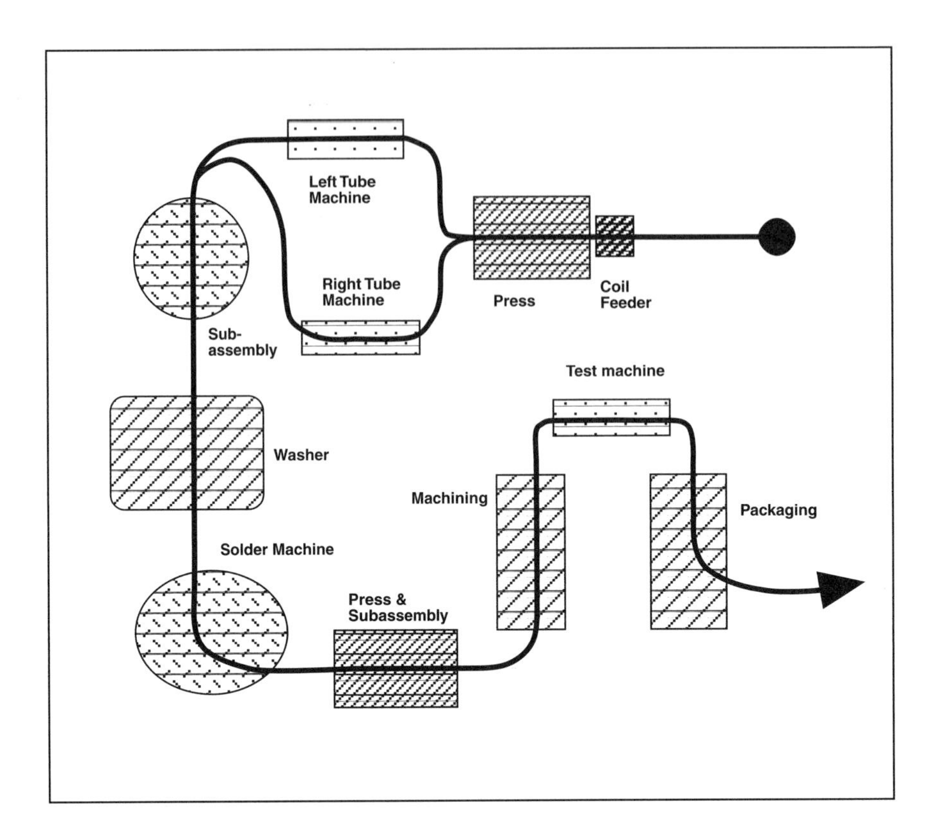

FIGURE 17-5

JIT U Line for Heat Exchanger Family at LNG

virtual line flow JIT production system at the top of the JIT region, slowly travel into the middle of the JIT region, and finally go to the physical line flow JIT system at the bottom of the region (see Figure 17-1). Three-step implementations have worked well in many organizations, although some, like Davidson, with manufacturing capabilities at or above the adult level, may require fewer steps. More than three steps are rarely required.

A THREE-STEP JOURNEY FOR IMPLEMENTING JIT[3]

A three-step journey for implementing JIT is described in this section. The steps are general, and the precise details will vary somewhat from company to company.

Step 1: Change the current batch flow production system to a *two-bin virtual line flow* JIT production system. Store inventory in bins on the plant floor in the areas where it will be used. Force improvements by lowering the amount of inventory in the bins.

Step 2: Change the two-bin system to a *kanban virtual line flow* JIT production system. Store inventory in "move" and "production" kanban in the areas where it is used and produced. Force improvements by lowering inventory—that is, reduce the number and size of the kanban.

Step 3: Create a *physical line flow* JIT production system by moving the equipment used in the kanban JIT system into a physical line flow. Again force improvements by lowering the inventory toward a target of zero.

These steps have the following characteristics:

- Only a few adjustments are made to the manufacturing levers at each step. The adjustments are simple, require little capital expenditure, and cause no disruption to ongoing production activities.
- Costs are reduced, and quality and delivery are improved at each step.
- Each step creates a foundation on which the next step is built.
- The JIT journey can stop at any step. Each step results in an effective JIT production system. Should the organization be satisfied with the improvements it has achieved or if other priorities arise, then the organization can pause or stop at that step.

Each step is described below in detail. In addition to outlining the mechanics of implementing a JIT production system, the descriptions will give an appreciation for what constitutes a JIT production system, how it is different from other production systems, and how it is much more than a collection of JIT techniques. This is important if we are to avoid the outcome experienced at LNG in Situation 17.2.

STEP 1: TWO-BIN VIRTUAL LINE FLOW JIT PRODUCTION SYSTEM

Each product or part in the two-bin JIT system is stored in a two-bin container that is located on the plant floor in the area where the part is used (see Figure 17-6). Documentation called a JIT package consists of a JIT card, visual control tag, manufacturing order, and part drawing, all in a clear plastic envelope. It is placed in the second bin of the two-bin container. The containers and the package are used in a pull production control system that works as follows:

1. As parts are required, they are taken from the first bin. When the first bin is empty, parts are taken from the second bin, whereupon the JIT package is taken from the second bin and placed in a nearby pickup area. The JIT package is now an authorization to produce a specified number of parts, the "replenishment quantity," within a specified number of days, the "lead time." This information is recorded on the JIT card.
2. A material handler collects the JIT packages from the pickup areas at regular intervals and delivers them to the manufacturing areas where the parts are produced. In the case of purchased parts, the JIT packages are delivered to a dispatching area. From there, they are sent to the suppliers.
3. Control boards are located in each manufacturing and dispatching area (see Figure 17-6). A control board is divided into six columns, one for each working day, Monday through Friday, and one overdue column. The columns are separated into weeks. When the material handler delivers a JIT package to a manufacturing or dispatching area, he or she takes the visual control tag from the JIT package and places it on the control board in a

FIGURE 17-6

The Two-Bin JIT Production System

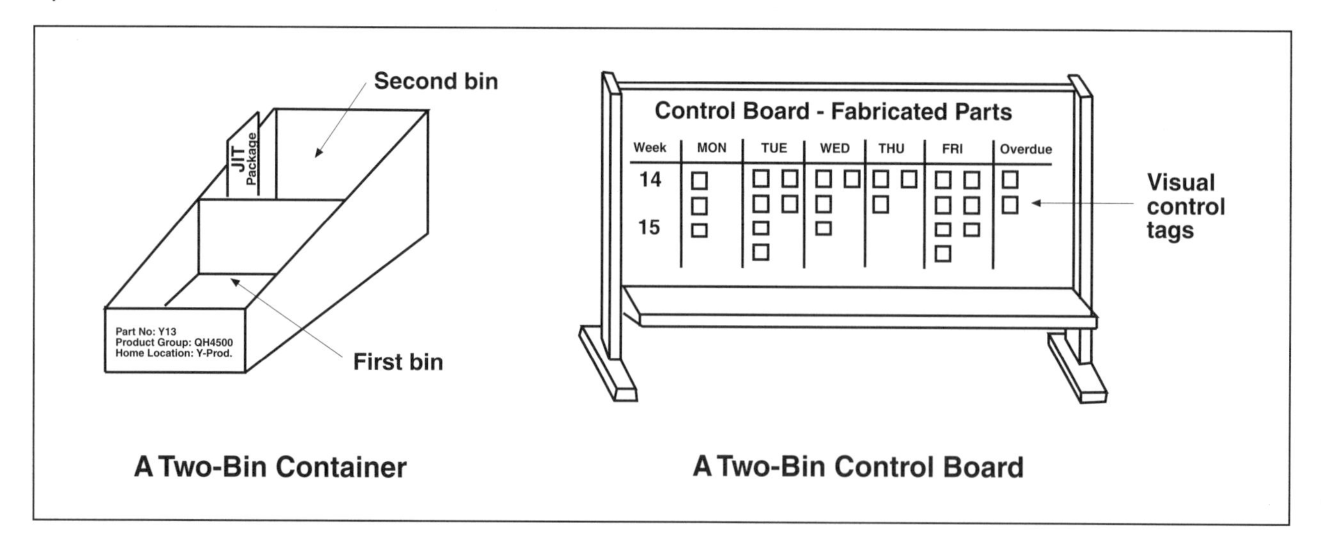

position corresponding to the week and day when the replenishment must be completed. For example, if the replenishment is triggered on Monday and the lead time on the JIT card is three days, then the visual control tag is placed in the Wednesday column, on the row corresponding to the current week. If the replenishment is not complete on Wednesday afternoon, the tag is moved to the overdue column, and action is taken to complete the replenishment quickly.

4. When the replenishment is complete, the visual control tag is removed from the control board and placed in the JIT package. The material handler delivers the parts and the JIT package to the department where the parts are used. There, the users/customers check that the order is complete and the quality is good. The parts are placed in the two-bin container. The second bin is filled to its specified level, and the remaining parts are placed in the first bin. The number of parts in the second bin is sufficient to meet the need during the lead time, including some safety stock. The replenishment quantity is a quantity of parts sufficient to satisfy a management-specified number of weeks of use.

The two-bin JIT system has a number of attractive features:

- Inventory is stored on the plant floor in the areas where it is used. This eliminates inventory transactions in and out of the stockroom. The stockroom can usually be reduced, and sometimes it can be eliminated altogether. This frees up floor space and reduces costs.
- Inventory is visible, making it easier to manage. Overproduction cannot occur because there is nowhere to put extra inventory.
- Production problems are visible. A perusal of the control boards shows what parts are late and which manufacturing areas and suppliers are busy.
- Each part is controlled by the area that uses it. That area triggers the replenishment, checks the quality, and holds the inventory.

- Only parts used regularly and with a moderate lead time can be controlled using the two-bin system. Parts not satisfying these criteria are controlled using the regular batch flow control system. These parts can be studied to see whether they can be standardized, replaced by other parts, produced in a different way, and so on, so that they can be controlled by the two-bin JIT system.

After the two-bin system has been working properly for some time, management reduces the amount of inventory in the bins in those areas of the plant where they wish to make improvements. Inventory is reduced by decreasing both the quantity in the second bin and the replenishment quantity. Replenishments then occur more frequently and in smaller quantities. Unless improvements are made to the production process, this reduction in inventory will cause shortages and production stoppages. The stoppages draw attention to problems and provide a motivation for implementing permanent solutions. As soon as a problem is identified, the inventory level is increased to its original level so that production can continue without further disruptions while personnel find and implement a permanent solution for the problem. When the solution is implemented, inventory is reduced again to expose the next problem. Making improvements to eliminate the problems exposed by the lowered inventories causes costs and cycle time to fall and quality to improve. This process of reducing inventory and making improvements is repeated again and again in all areas of the plant until all wastes have been eliminated.

SITUATION 17.3

Two-Bin JIT System for a Product Family

TWO PRODUCTS, m31 and m32, are assembled from 11 subassemblies and fabricated parts, x7, x12, x13, x17, y2, y7, y11, y13, z1, z3, and z5. All parts are to be produced using the two-bin virtual line flow JIT production system. The bills of material are shown in Figure 17-7. The two products, m31 and m32, have demands of 40 units per day each. The current plant layout is shown in Figure 17-8.

The initial design parameters for the two-bin JIT system are specified by management as follows. The replenishment quantity for each part will be a quantity

Product m31			Product m32		
Level	Part Number	Usage	Level	Part Number	Usage
0	m31	1	0	m32	1
..1	..x7	1	..1	..x7	1
....2	....y2	1	2	....y2	1
......3	......z1	3	3	......z1	3
......3	......z3	1	3	......z3	1
....2	....y7	1	2	....y7	1
......3	......z1	1	3	......z1	1
......3	......z3	1	3	......z3	1
......3	......z5	1	3	......z5	1
..1	..x12	1	..1	..x13	1
....2	....y2	1	2	....y7	1
......3	......z1	3	3	......z1	1
......3	......z3	1	3	......z3	1
....2	....y11	1	3	......z5	1
......3	......z1	2	2	....y11	1
......3	......z3	2	3	......z1	2
......3	......z5	1	3	......z3	2
			3	......z5	1
			2	....y13	1
			3	......z3	2
			3	......z5	2
			..1	..x17	1
			2	....y11	1
			3	......z1	2
			3	......z3	2
			3	......z5	1
			2	....y13	1
			3	......z3	2
			3	......z5	2

FIGURE 17–7

Bills of Material in Situation 17.3

sufficient to satisfy two weeks (10 days) of average use. The x-assembly area and the y- and z-production areas will replenish all orders within three days (that is, the lead time is three days). An amount of safety stock, equal to 50 percent of the average use over the lead time, will be added to the second-bin quantities to guard against shortages. Eleven bins are required. Four bins, for parts x7, x12, x13 and x17, are located at the beginning of the final assembly line because this is the area where they are used (see Figure 17-8). The four bins for y2, y7, y11, and y13 are located at the input area of the x-assembly area where they are used. Similarly, the bins for the z parts, z1, z3, and z5, are placed at the y-production area.

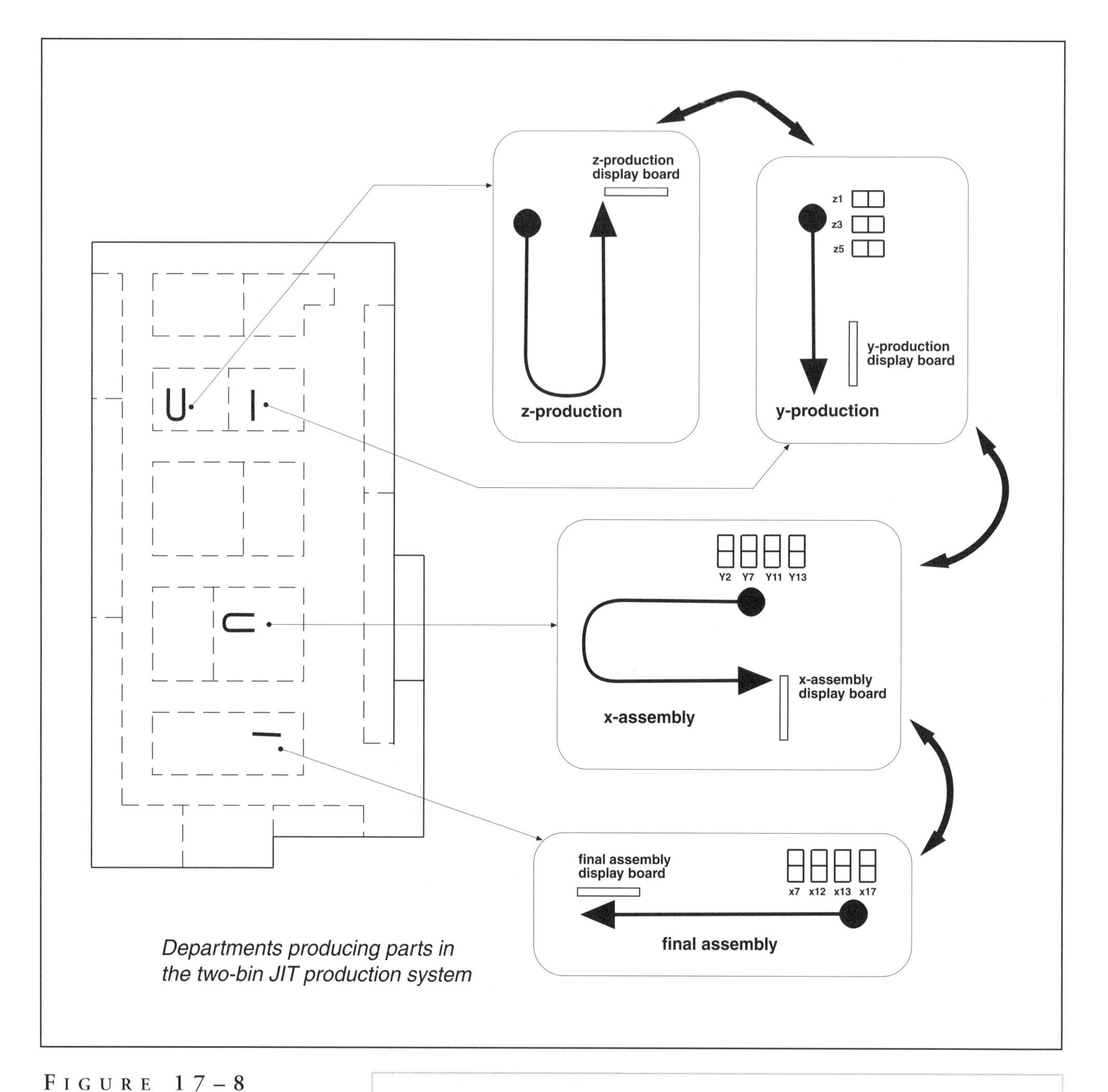

FIGURE 17–8

Layout under Two-Bin JIT in Situation 17.3

Figure 17-9 shows the calculations for the quantities in each of the bins. Consider, for example, part x7. Since one unit of x7 is required for each unit of product m31 and also for each unit of m32, 80 units of x7 will be required every day to meet the daily requirements for m31 and m32. Therefore, the replenishment quantity is 800 units (10 days at 80 units per day). The order point is 360 units (three-day lead time at 80 units per day, plus 50 percent for safety stock); that is, the second bin quantity is 360 units. Therefore, the two-bin container for part x7 must be large enough to hold 360 units in the second bin, and approximately 560 units in the first bin (calculated as the 800-unit order quantity less the expected use over the lead time—240 units). Inventory quantities for the other parts are calculated similarly.

It is also easy to determine quantities for parts used in both JIT and non-JIT products. In this case, the requirements for each part include the JIT requirements (as calculated in Figure 17-9) and the non-JIT requirements (as calculated by the batch flow production system).

The layout in the two-bin JIT system is shown in Figure 17-8. The only physical changes from the existing layout are the relocation of inventory from the stockroom to the plant floor areas where the JIT parts are used, and the additions of two-bin containers and control boards. These changes are easy to do, and require little if any capital expenditure. After employees become accustomed to the two-bin system and it has been operating properly for a time, management reduces inventory levels in the areas where they wish to make improvements. For example, suppose management is looking for improvements in the departments where y- and z-parts are produced. They might reduce replenishment quantities to one week's use and lead times to two days for these parts, and direct the y- and z-production areas to make improvements to their production processes so that they can produce these smaller quantities within the shortened lead time. The resulting inventory quantities are shown in the last two columns in Figure 17-9. It is easy to show that the average inventory levels for y- and z-parts are reduced by 46 percent. Under the two-bin JIT system, this routine of reducing inventory levels and making improvements continues for perhaps a year or two until no more improvements can be made.

Notice that production is intermittent and replenishment quantities are still relatively large under the two-bin

Part	Location	Parent parts	Requirements per day	Initial Quantities		Later Quantities	
				replenishment	second bin	replenishment	second bin
x7	final assembly	m31 m32	1 × 40 = 40 1 × 40 = 40 80	800	360	same	same
x12	final assembly	m31	1 × 40 = 40	400	180	same	same
x13	final assembly	m32	1 × 40 = 40	400	180	same	same
x17	final assembly	m32	1 × 40 = 40	400	180	same	same
y2	x assembly	x7 x12	1 × 80 = 80 1 × 40 = 40 120	1200	540	600	360
y7	x assembly	x7 x13	1 × 80 = 80 1 × 40 = 40 120	1200	540	600	360
y11	x assembly	x12 x13 x17	1 × 40 = 40 1 × 40 = 40 1 × 40 = 40 120	1200	540	600	360
y13	x assembly	x13 x17	1 × 40 = 40 1 × 40 = 40 80	800	360	400	240
z1	y assembly	y2 y7 y11	3 × 120 = 360 1 × 120 = 120 2 × 40 = 80 720	7200	3240	3600	2160
z3	y assembly	y2 y7 y11 y13	1 × 120 = 120 1 × 120 = 120 2 × 120 = 240 2 × 80 = 160 640	6400	2880	3200	2160
z5	y assembly	y7 y11 y13	1 × 120 = 120 1 × 120 = 120 2 × 80 = 160 400	4000	1800	2000	1200

Notes

1. Initial replenishment quantity is ten days' usage.
2. Initial second bin quantity is 150 percent of the usage over a three-day lead time.
3. Later replenishment quantity in departments y and z is five days' usage.
4. Later second bin quantity in departments y and z is 150 percent of the usage over a two-day lead time.

FIGURE 17–9

Two-Bin JIT System Calculations in Situation 17.3

JIT system. Consider, for example, part y7. Under the initial design parameters, a batch of 1,200 units is produced every 10 days. When inventory levels are reduced, batches of 600 units are produced every 5 days. Under JIT, the goal is to produce all parts on a continuous basis, which, in the case of y7, means producing 120 units each day or 15 units each hour (assuming eight hours are worked each day). This goal cannot be achieved with the two-bin system. However, the next step—the kanban JIT system—moves the organization closer to this goal. Two advantages of continuous production are the following: 1) Each part is produced on a regular basis, making it easy to study and improve the operations required to produce it. 2) It is possible to use smaller capacity equipment to produce the JIT parts (in the case of part y7, 15 units could be produced each hour on a small machine, compared to infrequent batches of 1,200 units for which a larger machine would likely be required).

This system is called a "virtual" rather than a "physical" line flow system because it behaves like a line even though it does not look like one. For example, the layout in Figure 17-8 is not a line layout. However, the two-bin containers, control boards, and the pull control system make the system behave like a line. We will see the same behavior in the kanban JIT system.

When as many improvements as possible have been made with the two-bin JIT system, the second step on the JIT journey can be taken. The kanban virtual line flow JIT production system can be implemented. This step is relatively easy because the two-bin system and the kanban system have many elements in common. In fact, the kanban system may be viewed as an *m*-bin system, where *m* is a number larger than 2.

Step 2: Kanban Virtual Line Flow JIT Production System

A kanban consists of a container of standard size in which a fixed number of units of a part are placed, and the paperwork that controls the use of the container and its contents. Two types of kanban are used; production kanban and move kanban. Production kanban are located at the end of the production area where the part is produced, and they do not leave that area. Move kanban are located at the start of the production area where the part is used. When a move kanban container is empty, it moves to the appropriate production area for replenishment. Kanban comes from the Japanese word for "card" or "ticket." Kanban paperwork can be as simple as a card specifying the part, the quantity in the container, and the location for the container (see Figure 17-10), or as detailed as the JIT package used in the two-bin JIT system. Paperwork is placed in a clear plastic sleeve on the side of the container. Kanban are used in a pull production control system that works as follows:

1. When the system starts for the first time, all production and move kanban containers are full.
2. Parts are taken from a move kanban container at the final assembly line and are used to produce final assemblies. When the last part is removed from a move kanban container the paperwork is taken from the container and placed on a kanban post (see Figure 17-10). The kanban is now called a move-free kanban. Production on the final assembly line continues, using parts from the next move kanban.
3. A material handler visits all kanban posts at regular intervals. He or she collects the move-free kanban and transports them to the production areas where the parts are produced, replenishes each empty move kanban container by taking parts from a full production kanban container, and then returns the full move kanban container to the area where the parts are used.
4. If a production kanban container is emptied while transferring parts from the production kanban to the move kanban, the material handler takes the paperwork from the production kanban container and places it on the kanban post. Now called a production-free kanban, the paperwork signals the production area to produce just enough parts to fill the empty production kanban container.

If there are too few move kanban, the area where the part is used could consume all the parts in the containers before a replen-

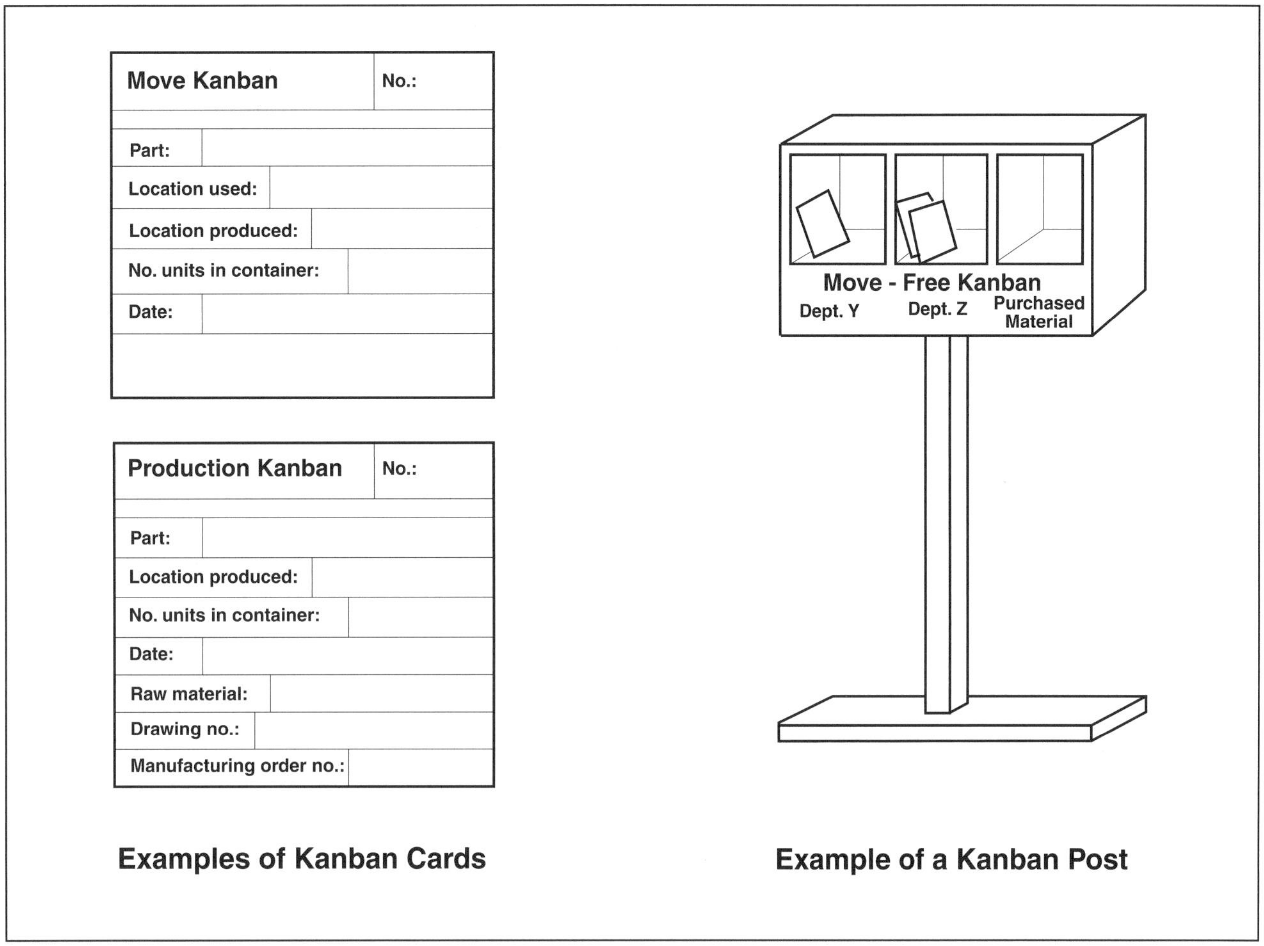
Move Kanban
No.:
Part:
Location used:
Location produced:
No. units in container:
Date:
Production Kanban
No.:
Part:
Location produced:
No. units in container:
Date:
Raw material:
Drawing no.:
Manufacturing order no.:
Examples of Kanban Cards
Move - Free Kanban
Dept. Y
Dept. Z
Purchased Material
Example of a Kanban Post

FIGURE 17–10

Elements in the Kanban JIT Production System

ishment can be made. If there are too few production kanban, empty move kanban arriving at the production area may not be replenished because more units are needed than are available in the production kanban. Either of these events can cause a stoppage at the area where the parts are used. The required number of move and production kanban are calculated so that there is just enough inventory to prevent these events from occuring.

d_i	Use rate for part i
$f_{i,m}$	Number of units of part i in a move kanban container
$f_{i,p}$	Number of units of part i in a production kanban container
p_i	Time required to produce $f_{i,p}$ units of part i
t_i	Transportation time for part i

The transportation time is the sum of the time to transfer an empty move kanban from the area where it is used to the area where it is replenished, the time to move parts from a production kanban to the move kanban, and the time to transfer the full move kanban back to the area from which it came. $f_{i,m}/d_i$ and $f_{i,p}/d_i$ are the times that one move kanban and one production kanban will last, usually between one and eight hours. Then, $t_i/(f_{i,m}/d_i)$ is the number of move kanban containers required to guard against shortages during the transportation time. Similarly, $p_i/(f_{i,p}/d_i)$ is the number of production kanban containers required to guard against shortages during the time it takes to replenish an empty kanban. It follows that the minimum numbers of move and production kanban containers for part i are:

$$k_{i,m} = \max(2, [t_i/(f_{i,m}/d_i) + 1]^+)$$

$$k_{i,p} = \max(k_{i,m}f_{i,m}/f_{i,p}, [p_i/(f_{i,p}/d_i) + 1]^+)$$

where $[x]^+$ denotes the smallest integer that is greater than or equal to x. These quantities provide just enough inventory to ensure that move kanban containers can always be replenished. Variations of the production control system, such as posting a move kanban when the first part is removed from a move kanban container, would result in slightly different versions of these equations.

Advantages of the kanban JIT system include those for the two-bin JIT system and some others:

- Production problems are visible. Large numbers of move-free kanban indicate problems with scheduling and material flow, while excessive numbers of production-free kanban indicate bottleneck workstations. Too few move-free and production-free kanban suggest that inventories are too high.
- The area that uses a part goes to the area that produces the part for frequent, small replenishments. This improves communication and provides feedback on quality.
- Each part is produced on a regular, continuous basis, making it easy to study the production operations and improve them.
- Smaller, less expensive equipment can be used.

After the kanban system is working properly, management reduces the amount of inventory by removing some move and production kanban, or reducing the number of units in a kanban, in those areas of the plant they wish to target for improvement. This strains the production system as it tries to operate with a smaller cushion of inventory. Soon problems occur that prevent replenishments before the remaining inventory runs out. Shortages and stoppages occur. These attract everyone's attention and are the motivation for making improvements. As soon as a problem is identified, the inventory level is increased to its original level so that production can continue without further disruptions while personnel find and implement a permanent solution. This process of reducing inventory and making improvements is repeated again and again in all areas of the plant until all wastes have been eliminated. The kanban JIT system allows management to fine-tune improvement efforts by, for example, reducing the number and size of kanban for certain parts at certain workstations so as to focus available resources onto the areas of greatest need.

SITUATION 17.4

Kanban JIT System for a Product Family

SUPPOSE the organization wishes to make further improvements beyond what was achieved with the two-bin JIT production system and decides to move to the kanban JIT system. Suppose also that the initial design parameters for the kanban JIT system are:

Production Area	$f_{i,m}, f_{i,p}$	p_i	t_i
x-assembly	10 units	8 hours	4 hours
y-production	20	8	4
z-production	40	8	4

One day consists of 8 hours.

The minimum number of kanban containers is calculated in Figure 17-11. Consider, for example, part x7. Eighty units per day, or $d = 10$ units per hour, are required to produce 40 units of m31 and 40 units of m32 each day. Because each kanban container contains $f = 10$ units, one kanban satisfies $f/d = 1$ hour of requirements.

The transportation time is t = 4 and so the minimum number of move kanban for x7 is max(2, [4/1 + 1]) = 5. The time required to produce a kanban of x7 is 8 hours, so the minimum number of production kanban for x7 is max(2 × 10/10, [8/8 + 1]) = 9. The number of kanban for the other parts are calculated similarly.

Small, inexpensive changes are required to the layout that was used in the two-bin JIT system. In the two-bin system, inventory was located in one area, the area where the part was used. Now inventory is located in two areas, the area where the part is used and the area where it is produced. Even though inventory is located in two areas, the total number of units in inventory is smaller than it was in the two-bin JIT system. For example, the maximum inventory for part x7 is 140 units (calculated from 5 + 9 kanban, with 10 units in each). In the two-bin system, the maximum inventory exceeded 800 units (the replenishment quantity for x7). The new layout is shown in Figure 17-12.

Once the kanban system is operating properly, management begins the process of reducing inventory for the purpose of forcing improvements to be made. For example, management may set the following targets for production times and move times:

Production Area	p_i	t_i
x-assembly	3.0 hours	2.0 hours
y-production	1.0	2.0
z-production	1.5	2.0

The reductions in the number of kanban are given in the last column of Figure 17-11. Significant improvements will need to be made if the production areas are to operate without shortages and stoppages with this reduced number of kanban.

Part	Location	Parent parts	Requirements per hour	No. of units in kanban	Initial No. of Kanban		Later No. of Kanban	
					move	production	move	production
x7	final assembly	m31 m32	1 × 5 = 5 1 × 5 = 5 10	10	5	9	3	4
x12	final assembly	m31	1 × 5 = 5	10	3	5	2	3
x13	final assembly	m32	1 × 5 = 5	10	3	5	2	3
x17	final assembly	m32	1 × 5 = 5	10	3	5	2	3
y2	x assembly	x7 x12	1 × 10 = 10 1 × 5 = 5 15	20	4	7	3	3
y7	x assembly	x7 x13	1 × 10 = 10 1 × 5 = 5 15	20	4	7	3	3
y11	x assembly	x12 x13 x17	1 × 5 = 5 1 × 5 = 5 1 × 5 = 5 15	20	4	7	3	3
y13	x assembly	x13 x17	1 × 5 = 5 1 × 5 = 5 10	20	3	5	2	2
z1	y assembly	y2 y7 y11	3 × 15 = 45 1 × 15 = 15 2 × 15 = 30 90	40	10	19	6	6
z3	y assembly	y2 y7 y11 y13	1 × 15 = 15 1 × 15 = 15 2 × 15 = 30 2 × 10 = 20 80	40	9	17	5	5
z5	y assembly	y7 y11 y13	1 × 15 = 15 1 × 15 = 15 2 × 10 = 20 50	40	6	11	4	4

Notes
1. $k_{i,m} = \max(2, [t_i/(f_{i,m}/d_i) + 1]^+) = \max(2, [4/(10/10) + 1]^+) = 5$
2. $k_{i,p} = \max(k_{i,m} f_{i,m}/f_{i,p}, [p_i/(f_{i,p}/d_i) + 1]^+) = \max(5 \times 10/10, [8/(10/10) + 1]^+) = 9$

FIGURE 17–11

Kanban JIT System Calculations in Situation 17.4

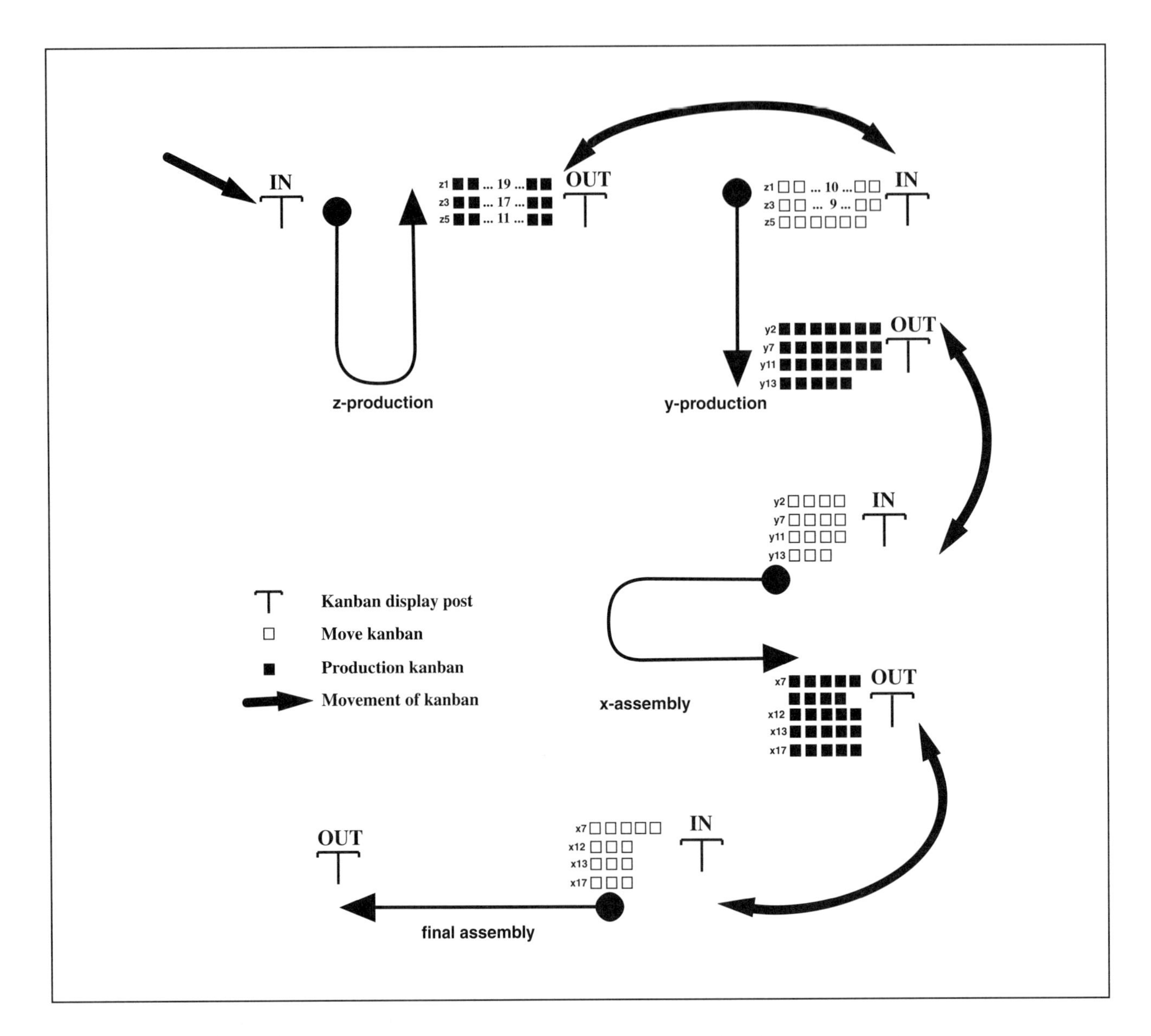

FIGURE 17-12

Plant Layout under Kanban JIT in Situation 17.4

The routine of removing kanban to force improvements to be made is repeated until no further improvement is possible. At this point, further improvement can be made only by rearranging the equipment layout into a physical line flow.

STEP 3: PHYSICAL LINE FLOW JIT PRODUCTION SYSTEM

Considerable improvements in cost, quality, and delivery have been achieved with the two-bin system and the kanban system. However, some organizations will require even lower costs, higher quality, and shorter delivery times and will move to the last step on the JIT journey—implemention of a physical line flow JIT system. Under this system, those elements of the production process that were so carefully improved under the two-bin and kanban systems are moved from their current locations to form a physical line.

The advantages of the physical line flow JIT production system include the following:

- Production of all parts is continuous.
- All operations required to complete the product are located in the same area, which improves communication and the ability to respond to problems.
- Capacities of all parts of the production process are balanced so that the entire process operates optimally.
- Inventory is reduced because of the elimination of transportation times. The amount of floor space used is also reduced.

SITUATION 17.5

Physical Line Flow JIT System for a Product Family

THE FINAL assembly line, x-assembly line, y-production area, and z-production area are studied to determine which equipment can be relocated into a special production line to produce products m31 and m32; assemblies x7, x12, x13, and x17; and parts y2, y7, y11, y13, z1, z3, and z5. The equipment is organized into a physical line flow—usually a U-line like the one shown in Figure 17-13. The four production processes (final assembly, x-assembly, y-production, and z-production) form one line, with the move and production kanban containers serving as links between the processes. Initially, the same number of kanban that were used with the kanban JIT system are used for the physical line flow JIT system. Because this is more inventory than is needed, the startup of the physical line flow system is easy. In the physical line flow JIT system, operators work inside the U-line and are trained to perform all the operations in the four processes. This makes scheduling easier, improves communication, and allows operators to help each other. U-lines require less floor space than traditional straight lines.

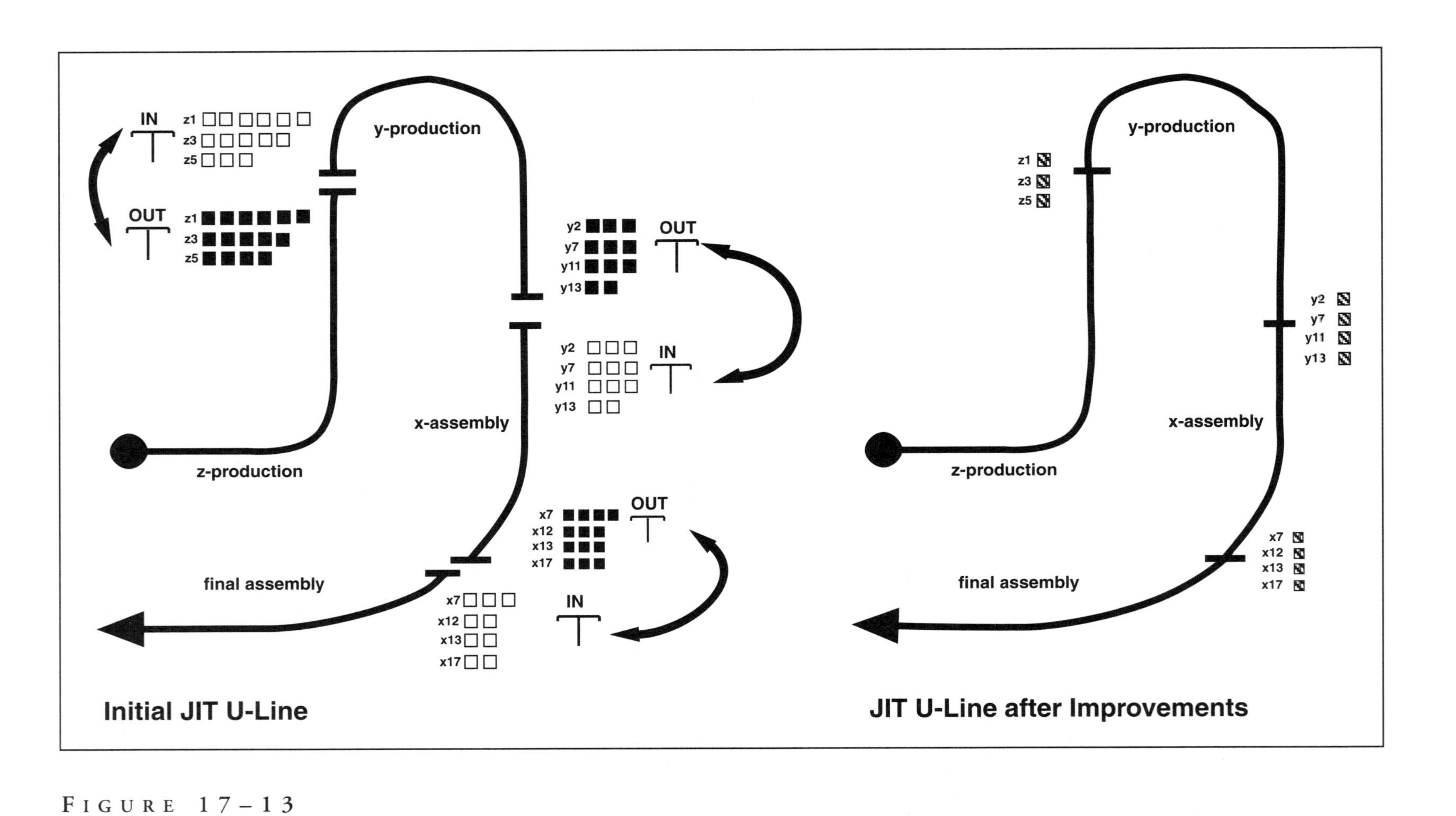

FIGURE 17–13

Plant Layout and Material Flow under the Physical Line Flow JIT Production System

After the physical line flow JIT system has been operating for a time, management gradually reduces inventories to identify opportunities for improvements. When problems occur that prevent replenishments from being made on time, solutions are found and implemented. This process continues until all problems or wastes have been eliminated. The capacities and production rates of the four production processes will be perfectly balanced and no inventory will be needed between the processes. The right side of Figure 17-13 shows the layout after numerous improvements have been made. When only one container, holding one unit, is needed between processes on the line, zero inventory has been achieved and the JIT journey ends.

The capital expenditure is the largest for this last step because equipment must be relocated. The disruption to production is small because the same equipment and kanban from the kanban JIT system are used.

The distinguishing feature of the JIT production system is its ability to identify and eliminate waste. When the end of the JIT journey is reached, all waste will have been eliminated from the production process. The resulting production system provides products at the lowest possible cost, with the best possible quality, in the shortest possible time. Few companies reach zero inventory. For most, it is a target to aim for, a motivation for making improvements.

Manufacturing Levers in the JIT Production System

The manufacturing levers are set so that the JIT mechanism for identifying and eliminating waste can operate to the fullest extent possible. At the same time, it is both a goal and a requirement that each lever have the highest possible level of manufacturing capability. Brief descriptions of each manufacturing lever follow. More information can be found in the Further Reading section at the end of this chapter.

Human Resources

The JIT production system is most efficient when it can obtain the most from both its people and its equipment. Operators are multiskilled. They are trained to operate several machines, participate in setups, help with maintenance, take part in problem solving, do housekeeping, perform quality control activities, and do material handling. There are only a few job classifications, and each job has a broad description. Managers are also multiskilled. They manage, provide training, facilitate problem solving, and make improvements. Operators and managers receive a lot of training. They are compensated according to the number of skills they have mastered rather than the specific task they may be doing.

Organization Structure and Controls

The JIT production system is usually a profit center. Line is more important than staff. The organization structure is flat and spans of control are wide. Decision making is pushed to the lowest level in the organization where a competent decision can be made. At the same time, the competence at all levels is improved continually through training. Extensive use is made of teams, especially for problem solving. Teams need not be self-managing. Organizations with a JIT production system are learning organizations because they are always making improvements. Benchmarking is used to set targets and identify new and better practices. Product engineering and process engineering work together in concurrent engineering programs to optimize product and process design simultaneously (see Chapter 10).

Control systems, such as compensation, individual and department performance measurement, accounting, capital appropriation, make versus buy, and supplier certification, are different from those used in other production systems. For example, traditional production systems use performance measures that track 1) equipment utilization; 2) ratios of direct labor, indirect labor, and overhead to volume; 3) number of setups, number of purchase orders; and so on. JIT performance measures are more direct. They track 1) actual cost and quality, 2) cycle time reduction, 3) delivery time and percentage on-time delivery, 4) actual production as a percentage of planned production, and so on.

Sourcing

The JIT production system requires a smaller number of suppliers compared to other production systems. These suppliers have high levels of manufacturing capability. A long-term partnership exists between the organization and its suppliers. Scheduling, design, cost, and other information flow easily between them. Suppliers are expected to make frequent deliveries of small lots, engage in problem solving and waste reduction, and participate in product design.

Production Planning and Control

The production planning and control lever is set so that the JIT mechanism for identifying and eliminating waste can operate to the fullest extent possible. Each area in the production process identifies its customers and suppliers. Each area produces only in response to pull signals—two-bin or kanban—from its customers. Inventory is located on the plant floor. Management carefully reduces inventory to strain selected areas for the purpose of forcing problems or wastes to be identified and eliminated. MRP is used for planning (capacity planning and setting a level master production schedule), and pull signals are used for execution and control of all production activities on the plant floor.

Process Technology

The JIT production system tries to get the most out of its processes and equipment. Much of the improvement activity in JIT focuses on improving equipment reliability and process capability (see Chapter 10). Because production is more continuous than in a batch flow production system, equipment can be smaller and can be run at a slower speed. Thus, equipment is often less expensive. Whenever possible, equipment is arranged into U-lines.

Short setup times are both a requirement and a goal in the JIT production system because many different products are produced in low to medium volumes. Efficient housekeeping and workplace organization are emphasized. Quality is very important. Statistical quality control techniques are used whenever possible for process control and problem solving.

Facilities

Facilities are medium size. The pace of production is steady but not as fast as in the equipment-paced or continuous flow production systems. Like process technology, facilities are changing constantly as improvements are made.

Manufacturing Outputs Provided by the JIT Production System

Cost and Quality

The JIT production system is a line flow production system. Consequently, it provides cost and quality at the levels associated with line flow production systems. JIT can do this because of its diligence at eliminating waste.

Performance

The JIT production system provides the same level of performance as the batch flow production system. Recall that both production systems are designed to produce many products in low to medium volumes. The volumes are not high, so the resources available to design new features and new processes are limited. This might seem surprising because companies that are well known for their JIT production systems, like Toyota, can provide high levels of performance. It is often the case that products manufactured by the JIT production system are so successful because the manufacturing outputs are provided at such high levels that demand increases and the production volume changes from low-to-medium to medium-to-high. When this happens, the virtual line changes quickly to a physical line and the production system changes to an operator-paced line flow system with a very high level of manufacturing capability. This operator-paced line flow system can provide the high level of performance. This is not unlike what happened at Samsung in Situation 11.1 (in Chapter 11).

Delivery

Delivery is provided at a high level because all the operations required to produce a product are done on equipment that is arranged along a virtual or physical line. Like all line flow production systems, little time passes between the time a product is started and is finished.

FLEXIBILITY AND INNOVATIVENESS

The JIT production system produces many products in low to medium volumes. In this respect, it is like the batch flow production system. Because it must produce many products, JIT provides flexibility and innovativeness at the same high levels as the batch flow production system. JIT can do this because of the high level of employee capability. The JIT production system can operate effectively only when the levels of manufacturing capability of the human resources and organization structure and controls levers are very high.

NOTES

1. Taiichi Ohno 1988, p. 97. (See the Further Reading section below.)
2. Adapted from C. Adair and E. Dawkins, "Workshop Report: Davidson Instrument Panel Division—Textron," *Target*, Vol. 3, No. 4, pp. 31–32, Winter 1987.
3. This section is based on work in J. Miltenburg and J. Wijngaard, "Designing and Phasing In Just-in-Time Production Systems," *International Journal of Production Research*, Vol. 29, No. 1, pp. 115–131, 1991.

FURTHER READING

Hall, R. W., *Zero Inventories*, Homewood, IL.: Dow Jones-Irwin, 1983.

Monden, Y., *Toyota Production System*, Second Edition, Atlanta, GA: Industrial Engineering Press, 1993.

Ohno, T., *Toyota Production System: Beyond Large Scale Production*, Portland, OR: Productivity Press, 1988. (Original Japanese edition published in 1978.)

Vollmann, T. E., W. L. Berry, and D. C. Whybark, *Manufacturing Planning and Control Systems*, Third Edition, Homewood, IL: Richard D. Irwin, 1992.

Wantuck, K., *Just-in-Time for America*, Milwaukee, WI: The Forum Ltd., p. 26, 1989.

Womack, J. P., D. T. Jones, and D. Roos, *The Machine That Changed the World: The Story of Lean Production*, New York: Harper Perennial, 1991.

CHAPTER 18

The Equipment-Paced Line Flow Production System

The equipment-paced line flow production system was developed by Henry Ford in the 1920s at the River Rouge plant in Detroit. Until that time, there were only two different production systems—job shop and batch flow. Ford's new system was a major advance. It provided much higher levels of the cost, quality, and delivery manufacturing outputs than the other two (see Figure 18-1). Before long, customer expectations for cost, quality, and delivery started to rise. Many organizations were forced to change their production systems to the new equipment-paced line flow system to meet these new expectations. For most, it was a difficult transition. We have had 70 years of experience with the equipment-paced line flow production system—time to learn and make improvements. Today, it is the easiest production system to design, manage, and operate.

Products and Volumes

The equipment-paced line flow production system is designed to produce several products in high volumes. Products are manufactured on specialized equipment using high volume tooling, all of which is arranged in a line flow (see Figure 18-1). Compared to the operator-paced line flow and just-in-time production systems, fewer products with higher, more regular volumes are produced by the equipment-paced system. The system produces standard products that compete on the basis of

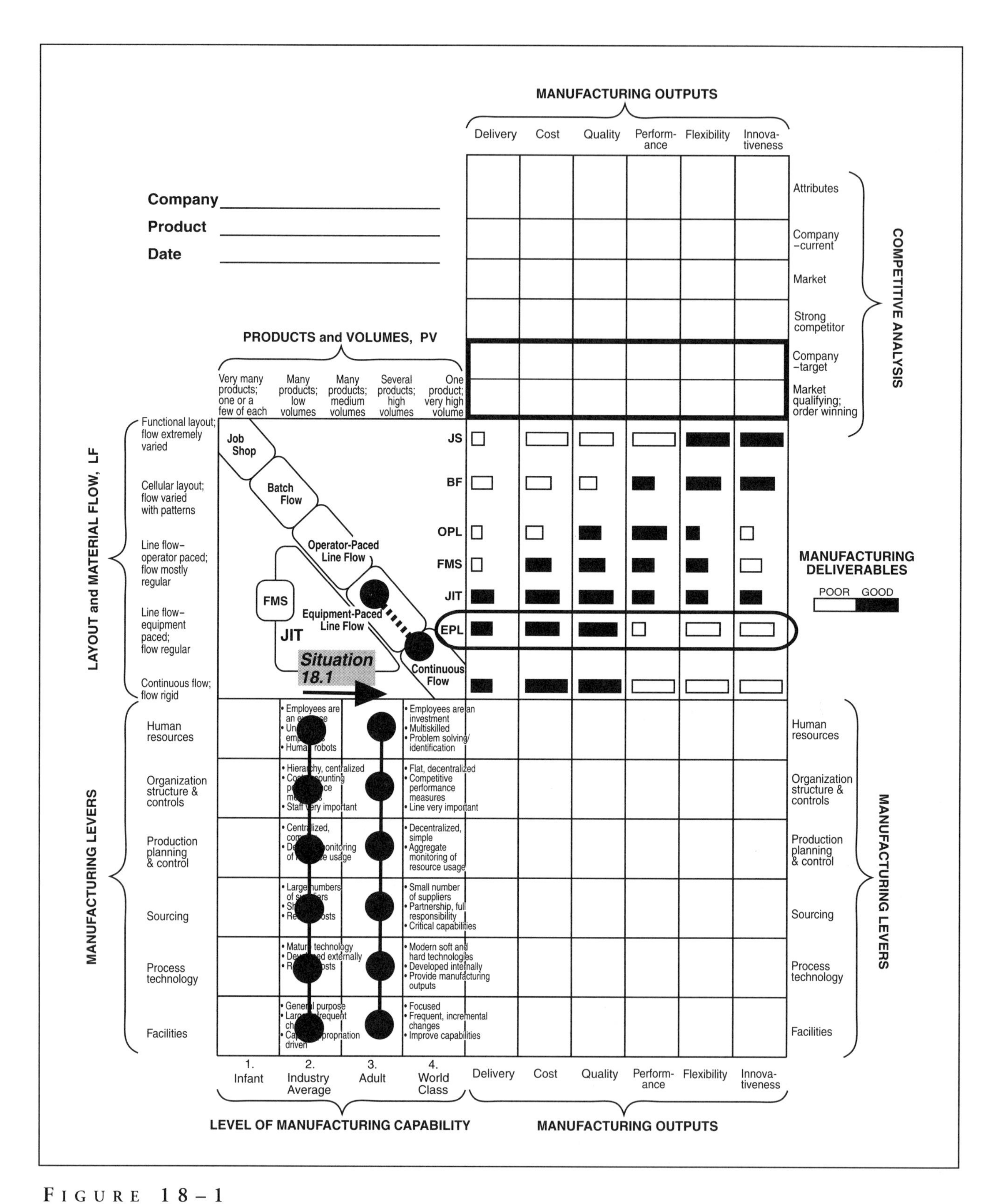

FIGURE 18–1

The Equipment-Paced Line Flow Production System

cost and quality. The amount of product change that can be accommodated is limited to a short list of options.

LAYOUT AND MATERIAL FLOW

A line layout is used in the equipment-paced line flow production system. A small family of products, each with almost identical manufacturing requirements, are produced on a common line. Equipment and tooling are highly specialized. They are designed to produce a few high-quality products quickly and, when the volume is high, at low cost. The equipment-paced line flow production system produces products at a rate equal to the line speed, which is the speed or pace at which the machines on the line operate. The rate cannot be varied. The line is either running or stopped. When it is running, a fixed number of operators are needed. If even one is missing, the line cannot run.

COMPETITIVE ADVANTAGE

It is almost a universal truth that the equipment-paced line flow production system is the best overall production system. It is the easiest system to design, manage, and operate. It provides high levels of the cost, quality, and delivery outputs. The equipment-paced line flow production system should be used whenever the following three conditions hold: 1) Customers want standard products with, at most, a small number of options (and these wants will not change in the near future); 2) the products can be produced in sufficiently high volumes over their design life to adequately utilize expensive, specialized equipment; and 3) customers desire high levels of the cost, quality, and delivery outputs.

MANUFACTURING LEVERS IN THE EQUIPMENT-PACED LINE FLOW PRODUCTION SYSTEM

HUMAN RESOURCES

Production operators are not as skilled as those in the job shop, batch flow, just-in-time, or operator-paced line flow production systems. The level of skill is comparable to the continuous flow production system. Detailed procedures and standards exist for all activities. They are necessary to keep costs low, achieve a high level of quality, and maintain a steady production rate on the line. Operators perform the same operations on each product. They are paid an hourly rate but a bonus scheme may be used to entice operators to come to work because any shortage of operators will stop the line. Numerous, large staff groups are needed to provide support services for production.

Organization Structure and Controls

The equipment-paced line flow production system is usually a cost center. Corporate influence on the production system in the areas of production scheduling, resource allocation, and capital expenditures is considerable. A centralized, bureaucratic organizational structure can best manage this high-volume business. The organizational structure is hierarchical, with many levels in the hierarchy. It is similar to the continuous flow production system. Staff groups are needed to manage product design, process engineering, production planning and control, maintenance, material handling, purchasing, quality control, and improvement activities. Close cooperation between product design engineers and process engineers is necessary to ensure that products are easy to produce on the specialized equipment. Even with close cooperation between these groups, making product design changes and introducing new products are difficult. Other important concerns of the staff groups include the following: Materials must be managed carefully so that the line is never stopped because of material shortages, careful planning is needed so that sufficient capacity is available to deal with customer orders without carrying excessive inventories or having to start and stop the line, and staff must keep current about developments in the process technology used on the line.

Sourcing

Because a few different products are produced, the equipment-paced line flow system requires a relatively small number of parts and materials in large volumes, at steady rates, over long periods, from a handful of suppliers. Consequently, the production system has considerable control over its suppliers. This results in supplier relationships characterized by long-term contracts, sharing of responsibilities for quality and improvements, participation in design activities, and so on.

Production Planning and Control

Production is most often scheduled to a forecast, with products allocated to a finished goods inventory. Schedules are planned sufficiently far in advance so that shortages are avoided. Like the continuous flow production system, forecasting, order scheduling, and material tracking are elaborate. Scheduling is not too flexible. The size of the purchased materials inventory depends on how much control the organization has over its suppliers. When its control is extensive, it can schedule purchased materials to arrive exactly when needed. When the control is limited, large inventories are carried so that the line does not stop because of shortages. Purchased material inventory is also high when there are economies of scale for ordering or transporting large quantities of materials. Work-in-process inventory is low because the line produces products quickly.

Process Technology

Equipment, fixtures, and tooling are specialized. There is a considerable amount of automation. However, there is less specialization and automation than in the continuous flow production system because several products (or one product with several options) are produced on the same line, compared to only one product in the continuous flow production system. The time required to change from one product to another is sufficiently long that lengthy runs of each product are scheduled. This minimizes the production time lost because of setups and reduces quality problems that occur when changing from one product to another. Organizations with an equipment-paced line flow production system have setup time reduction programs to make shorter production runs possible.

Because most of the value added is provided by machines, process technology is important. Consequently, the equipment-paced line flow production system is a technology leader. An important part of the process engineering department's job is to keep current on new developments in process technology. Changes in technology are usually major and occur infrequently. When they do occur, extensive changes are made to the line. This is also true for the FMS and continuous flow production systems. (Operator-paced, just-in-time, batch flow, and job shop production systems are different: Process technology changes are small and occur more frequently.) The equipment-paced line flow production system is capital intensive. Hard technologies are more important than soft technologies. Indirect costs, such as staff, are high. Of the direct costs, material cost is high and direct labor is low.

Facilities

Facilities are large and seek to achieve economies of scale. The process is speedy. The equipment on the line is synchronized, so there are no bottlenecks. Equipment utilization and labor efficiency are high.

Manufacturing Outputs Provided by the Equipment-Paced Line Flow Production System

Cost and Quality

The equipment-paced line flow production system is designed to produce high volumes of a few products. High-volume tooling is used. Machines, fixtures, and operators are all specialized. The results are low cost and high-quality products.

Performance

The equipment-paced line flow production system can provide only an average level of performance for two reasons. On the positive side, the high volume of each product makes it possible to invest in research and development to design a steady stream of new features and to improve existing processes so that current products can be produced more efficiently. On the negative side, the high cost of changing specialized tooling and equipment and of training relatively unskilled operators make the changes needed to provide a high level of performance year after year quite difficult.

Delivery

Machines are specialized and synchronized so that products can be produced at a steady, fast rate. Operators perform relatively simple tasks that are coordinated carefully with machine operations. The result is fast and reliable delivery.

Flexibility and Innovativeness

The levels of flexibility and innovativeness provided by the equipment-paced line flow production system are low for precisely the same reasons that the levels of the cost and quality outputs are high. The highly specialized machines and tooling and relatively unskilled operators make it both expensive and difficult to change the existing product mix, make design changes, and introduce new products. Even small design changes require expensive tooling alterations. Introducing new products often requires a shutdown of one or more weeks.

SITUATION 18.1

Whirlpool Changes an Equipment-Paced Line Flow Production System

IN 1980, Whirlpool Corporation, with annual sales of over $4 billion, found that its washing machines were overpriced. Whirlpool's basic model was produced on an equipment-paced line flow production system. However, the production system, which was designed in the 1950s, could no longer provide market qualifying levels of the cost and quality outputs. Three changes had to be made: 1) The production system was moved closer to a continuous flow production system (see Figure 18-1), which meant that less labor and more automation would be used, and volume would be increased; 2) the product was redesigned to make it easier to manufacture on the new, highly automated equipment-paced line; and 3) numerous adjustments to the manufacturing levers were made to raise the manufacturing capability from an industry average to an adult level. Practices and policies in human resources were changed. Production support departments and reward systems were also changed. Improvements were made at suppliers. Production planning and control activities were improved. Several years and $162 million later, the new equipment-paced line flow production system was the largest producer of washing machines in the world. Costs were lower, quality was higher, and volume was up 80 percent.

SITUATION 18.2

Television Manufacturer Fine-Tunes Production System

PHL ELECTRONICS produced 19-inch through 52-inch television sets at a plant in Europe. The plant was organized into fabrication, subassembly, and final assembly departments. A batch flow production system was used in the fabrication area, an operator-paced line flow production system was used in the subassembly department, and an equipment-paced line flow production system was used in the final assembly department. Many just-in-time techniques, including setup time reduction, improved housekeeping, and improved supplier relations, had been implemented recently in the fabrication and subassembly departments. Improvements had also been made in the final assembly department, where new bar-code and scheduling systems had been installed.

In the following discussion, the manufacturing levers in the final assembly department's equipment-paced line flow production system are examined. The final assembly department included two assembly lines. The smaller and slower of the two lines ran one shift a day, five days per week, and produced the largest televisions. The other line was newer and faster and produced regular size televisions, two shifts a day, five days per week.

Human Resources in Final Assembly

Most assembly line operators were trained only to work at one station. The few who were trained for work at more than one station were used as relief operators, moving from station to station to relieve operators for short breaks. The lines moved at a constant speed, and operators had a fixed amount of time to complete their tasks. Quality control tasks were done by quality control personnel who took regular measurements (for SPC charts) at special stations along the line. Maintenance personnel performed maintenance and repair activities. Material handling personnel brought parts to the line and took away scrap and empty containers. All employees received an hourly wage. There was no incentive wage plan.

Organization Structure and Controls in Final Assembly

The assembly department was a cost center. Management was evaluated on the number of units produced, equipment utilization, labor efficiency, and actual costs compared to budget or standard costs. Every effort was made to keep the lines running by avoiding breakdowns and minimizing the number of times lines stopped to change from one size and model of television to another.

The organizational structure was hierarchical. There were many staff groups, including process engineering, material handling, production planning, maintenance, quality control, and purchasing. Each staff group reported to its own general manager—who also managed similar groups in other parts of the company—and to the general manager of the assembly department.

Sourcing in Final Assembly

High volumes of standard raw materials and parts were purchased from other departments and from outside suppliers. A large purchasing department managed this activity for the assembly department and for all other departments in the company. Suppliers were selected on the basis of low cost, satisfactory quality, and consistent delivery. The assembly department had considerable influence on its outside suppliers. It had much less influence on the other departments in the company who supplied it with parts.

Production Planning and Control in Final Assembly

Production scheduling for the two assembly lines was done by production planners in close consultation with the marketing department. A sophisticated MRP II system was used. Quantities and delivery times for purchased parts were planned far into the future, and the resulting schedules were conveyed electronically to suppliers. Inventories of raw materials and purchased parts were small. Work-in-process inventory and finished goods inventory were also small.

Recently, the assembly department, in partnership with the distribution group, implemented a bar-code control system. The new system provided timely information for planning and control activities. Self-adhering bar-code labels were attached to key parts of the television as it moved down the line, beginning with the picture tube. Other labels were placed on the inside and outside of the cabinet and on the cartons for warehousing and shipping.

Fixed electronic readers scanned the bar code as the television moved down the assembly line. The status of each television was known at all times, even when it was taken off the line for rework. No television could leave the assembly department for the finished goods warehouse or shipping until it had been accounted for by the computer. When the television passed the final bar-code reading station, the computer system released all the parts necessary to build another set from inventory. Known as *backflushing,* this technique helped control inventories in the assembly department and at the suppliers. Portable hand-held readers were used to scan the bar codes in the shipping area. The result was an accurate record of finished goods inventory and the orders that had been shipped. Three peel-away labels were placed inside the television; these could be used later by service representatives if the set had to be repaired in the field.

Process Technology in Final Assembly

The equipment and tooling used on the lines were very specialized and expensive. The time required to change from one size and model of television to another was not excessively long, but production schedules still emphasized long runs of each product to minimize the production time lost because of changeovers. The process engineering group was always on the lookout for new process technology that could produce televisions faster, at lower cost, or with higher quality. The group also worked closely with the design engineering group to ensure that new products and design changes to existing products were done so that the products would be easy to produce.

Facilities in Final Assembly

The assembly department was large, modern, and clean. It was designed to achieve economies of scale. Televisions were produced at a quick rate. There were no bottlenecks. The equipment utilizations and labor efficiencies were high.

CHAPTER 19

The Continuous Flow Production System

Many producers of metals, plastics, chemicals, pharmaceuticals, and lumber use continuous flow production systems to produce high volumes of these standard products. For most of these producers, "the process is the product." The process adds all the value; if something goes wrong with the process, there is no product. For example, if a grade of wire does not meet all its specifications, it cannot be sold as that grade of wire. The "conformance to specifications" aspect of quality is crucial. Plants that use continuous flow production systems are highly automated and expensive.

Products and Volumes

The continuous flow production system is designed to produce high volumes of one product or a small product family on highly automated, specialized equipment (see Figure 19-1). Production is continuous—usually 24 hours a day, seven days a week. Most often the continuous flow production system is used to produce commodities. Commodities are mature products whose designs or specifications are standardized throughout the industry. Because all competitors produce identical products, cost becomes the basis of competition. Product design changes and new product introductions occur infrequently.

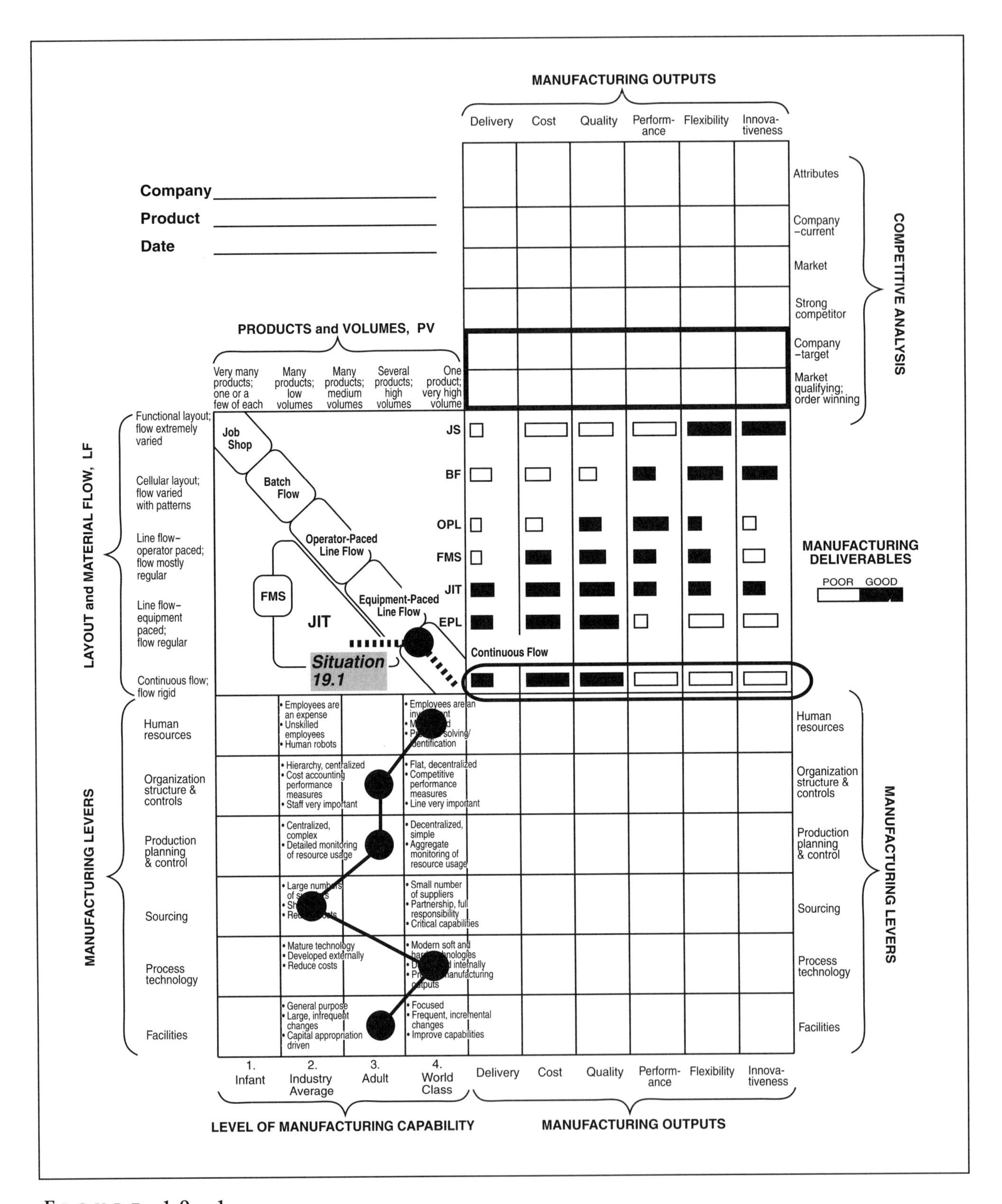

FIGURE 19–1

The Continuous Flow Production System

Layout and Material Flow

A line layout is used in the continuous flow production system. One product or a small family of products, each with identical manufacturing requirements, is produced on a highly automated line. Equipment and tooling are very specialized. A few operators are needed to monitor the equipment on the line. Products are produced continuously at a rate equal to the line speed, which cannot be varied. The line is either running or it is stopped. The products are produced quickly, with very high quality and at low cost.

Competitive Advantage

The continuous flow production system is used when three conditions are satisfied: 1) Customers require the best possible cost, quality, and delivery; 2) the volume of the product is sufficiently high to keep an expensive, highly automated, specialized line running continuously; and 3) the product design is stable. These conditions are satisfied when a product is in the mature stage of its product life cycle (see Chapter 11).

Manufacturing Levers in the Continuous Flow Production System

Human Resources

Production operators are relatively unskilled compared to those in other production systems. They are responsible for monitoring the highly automated production process to ensure that equipment is operating within prescribed parameters. Operators track quality attributes on SPC charts and perform routine maintenance and housekeeping tasks. All employees are paid an hourly rate. No incentive pay scheme is used, although bonuses may be awarded when record levels of production output are achieved. Maintenance is a critical activity. Maintenance personnel are responsible for preventive maintenance, maintaining spare parts, and repairs when breakdowns occur. A breakdown anywhere in the process brings the entire process to a halt.

Organization Structure and Controls

The continuous flow production system is usually a cost center. The high volume and rigid nature of the production system lends itself to a centralized, bureaucratic organizational structure. The structure is hierarchical, with many levels. Corporate influence on the plant in the areas of production scheduling, allocation of staff resources, and capital expenditures is high. Management is evaluated on the number of units produced, equipment uptime, and actual costs compared to budget or standard costs. Every effort is made to keep the process running. Changes to the production process are expensive and require that production be stopped. A capital appropriation request to the corporation precedes these changes. The request is accepted or rejected on the basis of projected cost savings.

The continuous flow production system requires numerous large staff departments to perform activities such as planning and scheduling production and maintenance, forecasting raw material requirements, tracking materials, making process improvements, and managing quality activities. Staff departments are often more influential than line departments. A separate quality control department is responsible for all aspects of quality in the production system. This includes SPC activities in the production process, quality of incoming raw materials, and projects for improving quality.

Sourcing

The continuous flow production system requires large volumes of a few raw materials, many of which are commodities. Consequently, it has a great deal of influence over its suppliers. Suppliers are selected on the basis of low cost, so long as quality is high and delivery is reliable. Information on raw material requirements—quantities and delivery times—is known well in advance and is sent electronically to the suppliers through the computerized production planning and control system.

Production Planning and Control

Production is make-to-stock; that is, production is scheduled according to a forecast, with products allocated to a large finished goods inventory. Customer orders are filled from this inventory. Work-in-process inventory is low. Raw material inventory is high so that quantity discounts can be realized and shortages (which would stop the production process) do not occur. Production schedules are planned far into the future. Changes to production schedules can be made, but there is usually a long period during which the schedule is frozen. Sophisticated computerized systems are used to plan, control, and optimize all activities in the production system.

Process Technology

Equipment is highly specialized, tooling is high volume, and the process is highly automated. Everything is expensive. Changeover times are long. Consequently, one product, or a small family of similar products, is produced. In the latter case, production runs are very long to minimize changeovers. SPC is used extensively. Control charts and other SPC techniques are easy to apply when a single product is produced continuously, and they are effective at maintaining high levels of quality throughout the process.

Because all the value added is provided by machines, process technology is very important. The continuous flow production system is usually a technology leader. An important part of the process engineering department's job is to stay current with new developments in process technology. Technological change is revolutionary rather than incremental and takes the form of new, faster equipment and processes. Additions to capacity can be made only in large pieces. Continuous flow production systems are capital intensive. Indirect costs, such as staff, are high. Of the direct costs, material cost is high and direct labor is low.

Facilities

Facilities are large to achieve economies of scale. The pace of production is very fast. There is no bottleneck machine because all machines have the same capacity and run at the same speed. The continuous flow production system is large and produces high volumes of output, so it also produces large amounts of waste, some of which is hazardous. Thus, the production system forces the organization to spend considerable resources to handle and treat waste.

Manufacturing Outputs Provided by the Continuous Flow Production System

Cost and Quality

The continuous flow production system is designed to produce one product or product family in very high volumes. Machines and fixtures are specialized. High-volume tooling is used. The process is highly automated and although this is expensive, the volume is high enough that the process can be run continuously—often 24 hours a day, seven days a week. All this results in products with the lowest possible cost and the highest possible quality.

Performance

The continuous flow production system cannot consistently provide a high level of performance. A high level of performance requires a constant stream of product design changes, new products, and process changes. This high level is difficult to achieve because the production system is so rigid. It is very costly to change the specialized, automated equipment; to retrain the relatively unskilled operators; to stop production for a few weeks to make the changes; and so on. These steps can be done from time to time, but not at the pace needed to provide a high level of performance year after year.

Delivery

The continuous flow production system is highly automated. Machines perform all the operations; operators simply monitor the process. The process is carefully designed to produce only one product or product family and to produce it continuously. The result is the fastest possible delivery time. The delivery time reliability is also very high, provided maintenance is done well and suppliers are well managed.

Flexibility and Innovativeness

Flexibility and innovativeness are difficult to provide at high levels. The specialized machines are designed to run at one speed that cannot be altered. Changing the design of a product or introducing a new product usually requires extensive changes to the specialized machines and tooling, retraining of staff and unskilled operators, and a long shutdown.

SITUATION 19.1

A Manufacturing Strategy for the MGR Steel Company

PROBLEM[1]

The MGR steel plant was one of four plants in the northwest. Each plant used minimill technology consisting of an electric arc furnace, a continuous caster, and a rolling mill, all operating as a continuous flow production system. MGR produced two product families: lightweight steel bars for reinforcing concrete and lightweight structural shapes such as angles, rounds, and flats. The MGR plant was the newest plant. When it was built, the capacity of the rolling mill (500,000 tons per year) was twice that of the caster and furnace (250,000 tons per year). The plan was to add a second furnace and caster to balance the plant once it became profitable. Five years later, after it had captured one-third of its local market, MGR expanded its furnace and caster capacity to 320,000 tons per year and began selling the product outside the local market. MGR thus avoided capturing too large a market share, which could depress prices and reduce profit margins.

Process Technology at MGR

Furnace. Shredded scrap metal was picked up by an overhead crane and loaded into a charging bucket, from which it was charged (or loaded) into the electric arc furnace. Other ingredients such as lime and alloys were added to the charge. Graphite electrodes were lowered into the furnace. When the power was turned on, an electric arc formed between the electrodes and the scrap, and the resulting intense heat melted the scrap quickly.

Continuous caster. Once the batch of steel in the furnace, called a heat, was melted completely, the furnace tilted and metal poured into a ladle. The full ladle was lifted by the ladle crane to the top of a continuous casting machine. The molten metal was released through a hole in the bottom of the ladle into a rectangular, trough-shaped dish called a tundish. There were four holes in the bottom of the tundish, one for each strand of steel the machine cast. Metal flowed through the holes into molds. The molds oscillated to produce homogenous, billet-shaped strands, approximately 5″ × 5″ in cross-section. The strands flowed from the molds into spray chambers. Beyond the spray chambers, the strands solidified, were cut by an automatic torch into standard lengths called billets, and moved through straighteners. The last step was air-cooling on racks that moved and rotated each billet along a long table.

The industry standard time to cast a 100-ton heat from the furnace was about two hours. The tundish and molds were cleaned after each heat, which took 12 minutes. At the same time, the ladle crane lifted the empty ladle from above the tundish and lowered it to the ground. The crane then picked up a full ladle that had just been tapped and raised it up to the caster, thereby beginning the cycle again.

Rolling mill. Before being rolled into finished shapes, the billets were reheated in a gas furnace. They were released individually from the furnace, and they traveled through 15 rolling machines. These machines were sets of heavy rollers that used pressure to squeeze the hot billet into smaller and smaller sizes until it reached its final shape. The bars cooled on long racks and then were cut and bound automatically for shipping.

Changing from one product to another required changing some of the rollers in the rolling machines. The number of rollers that needed to be changed depended on whether the next product was a different shape or whether it was simply a different size of the product that was currently being rolled. The time to complete a changeover varied from 30 minutes to four hours.

Human Resources at MGR

MGR's strength was its people. The company practiced participatory management. It believed that decisions should be made at the lowest competent level in the organization, and that the lowest level should be as competent as possible. An extensive training program was in place to support this belief. The company had never laid off an employee, but it had fired a few who did not "fit in." The company also had a significant profit-sharing plan.

Organization Structure and Controls at MGR

Meetings and memos were discouraged in favor of phone calls and face-to-face conversations. Everyone was on a first-name basis and people were free to speak their minds. The plant was designed to support this approach. Offices and conference rooms were spread throughout the plant. The management structure was lean and flat, with only four levels from top to bottom. Part of the MGR culture was that staff personnel served operations first and foremost.

Production Planning and Control at MGR

An important objective in production planning and control was to synchronize the caster with the furnace. There were delays on the caster from the outset. Initially, the life of the tundish was only four or five heats. Employees were able to increase the life to eleven heats by experimenting with different brick liners. Planners tried to minimize the time spent doing changeovers in the rolling mill. Consequently, long runs of each product size and shape were scheduled. The nozzles in the spray chambers had to be changed every three heats. Sometimes one of the four strands would clog or the steel would break out, and the strand had to be plugged off, cleaned, and restarted after the heat was completed. This resulted in a slower production rate because, with fewer strands open, the molten metal emptied from the tundish at a slower rate.

Level of Manufacturing Capability at MGR

MGR had an adult level of manufacturing capability (see Figure 19-1). The company expected its employees to make improvements. For example, the following four inexpensive improvements increased the size of a heat from 100 tons every two hours to 120 tons every 90 minutes: 1) More cooling was added to the electrical transformer, 2) the welds on the furnace shell were reinforced, 3) a new type of refractory brick was used to line the furnace, and 4) the size of the ladles was increased by adding two 12-inch rings to the top of each ladle.

Six years after the plant was built, improvements increased the capacity of the electric arc furnace from 250,000 tons per year to 320,000 tons. New technology, developed in-house, promised an additional increase in capacity to 400,000 tons within a few years. While this increase posed no problem for the rolling mill, rated at 500,000 tons per year, the continuous caster, originally rated at 250,000 tons per year, would become a severe bottleneck. MGR was now considering four different proposals for increasing capacity and further improving its production system.

Proposal 1. A team had just returned from Japan, where it examined a virtually nonstop version of continuous casting that could eliminate the bottleneck. By carefully synchronizing existing equipment and adding one new piece of equipment, the Japanese had developed a casting procedure that permitted many heats to be cast consecutively, without stops to clean out the tundish and molds. Casting was interrupted only when the refractory brick in the tundish wore out. The key piece of additional equipment was a ladle car platform above the caster. The platform consisted of a track with three ladle car positions—one over the tundish and one on each side. While metal from one ladle flowed into the tundish, the furnace was tapped into another ladle. The ladle was raised to one of the empty ladle car positions. The full ladle reached the platform in time to slide over the tundish when needed, but not so early that the temperature in the ladle dropped below the casting temperature. When the ladle over the tundish was empty, it was moved into the empty slot and the full ladle was rolled into place.

Under optimum running conditions, this system permitted long runs of one product using steel with identical metallurgy. It also reduced scrap (normally produced at the beginning and end of each heat). MGR estimated that its yield would increase from 88 percent to 95 percent. The new technology would be expensive and would require four to six weeks for implementation, during which time the plant would be shut down.

Proposal 2. The chief engineer had a different idea. Instead of using more ladles and a ladle car, make better use of the crane to synchronize the furnace and the caster. The crane would lift a full ladle up to the tundish, let it empty, bring it down, tap into another ladle, then bring it up. The trick would be to do all the crane work before the tundish emptied. The goal was to make the continuous flow production system even more continuous. The production superintendent disagreed. There was no precedent for this kind of activity. Hurrying a crane could be dangerous and, if there was a mistake in timing, the company could end up with a ladle filled with solid steel. Given that the tundish emptied out in less than four minutes, he did not think the chief engineer's idea would work.

Proposal 3. The comptroller and marketing manager also had an idea. Their reading of the market suggested that the plant ought to increase the variety of shapes and sizes it produced and develop new markets in which to sell them. The two managers felt that the construction boom in the area was over, at least temporarily, and so the demand for lightweight steel bar would be down. They felt that the demand for other structural shapes would continue to grow. The managers knew that there were some problems with this approach. Though demand for the group of products had grown, the demand for any one of the products was hard to predict. And the metallurgy for each product was somewhat different. These problems would have to be solved or MGR would end up with large inventories of the wrong products.

Proposal 4. The materials manager was worried about the supply of raw materials. The price of scrap had fluctuated wildly during the last year. Some minimills had purchased scrap yards to secure supply. Perhaps MGR should do this.

Solution

MGR used the continuous flow production system to produce a few products in large volumes (see Figure 19-1). MGR had a high level of manufacturing capability—each lever was at or above the adult level—as a result of its improvement activities, especially in the human resources and process technology areas.

In any continuous process production system, the capacity of all process segments should be approximately the same. Since the lowest capacity segment determines how much will be produced, any process segment whose capacity exceeds this will have unused capacity. Unused capacity results in increased costs because of the capital costs associated with purchasing the capacity and constructing the facilities in which it is housed, and the operating costs associated with maintaining it. Because continuous flow production systems compete on cost, there is a strong incentive to use the capacity or eliminate it. Consequently, it is reasonable for MGR to want to increase the capacity of the continuous caster to balance it with the furnace and the rolling mill. The problem is to decide what kind of capacity to acquire.

Implementing proposal 1, the Japanese continuous casting procedure, represents movement down the diagonal of the PV–LF matrix (refer to Figure 19-1). The plant layout/material flow will be more rigid and continuous because the link among the three process segments will be tighter. The number of products does not change, but the production volumes increase. Proposal 1 is quite appropriate for MGR. Proposal 2 would move MGR even further down the diagonal; it would make the process even more continuous. It cannot be adopted, however, because it does not appear to be feasible for the products MGR produces. Proposal 3 would broaden the product mix by producing smaller quantities of more products. This represents a horizontal movement to the left on the PV–LF matrix, which would be difficult even for a continuous flow production system with a very high level of manufacturing capability. MGR should not implement this proposal unless it is prepared to change its production system to an equipment-paced line flow production system and change its manufacturing strategy.

The final proposal focuses on the vulnerability of the continuous flow production system to disruptions in the supply of raw materials. Regardless of what is done with the other proposals, MGR must ensure that it has a reliable supply of high-quality, low-cost raw materials. Proposal 4, or some variation of it, should be implemented.

NOTES

1. Adapted from K. Clark, "Chaparral Steel," Harvard Business School Case, Number 9-687-045, and R. Radford and P. Coughlan, "Steel Industry Technical Note," University of Western Ontario, Case Study Technical Note, Number 986D013.

About the Author

John Miltenburg is professor of production and management science in the Michael DeGroote School of Business at McMaster University in Hamilton, Ontario, Canada. Before joining McMaster University Dr. Miltenburg worked for General Motors. In 1988 and 1989 he was a visiting professor at the Eindhoven University of Technology in the Netherlands.

Dr. Miltenburg has published more than 30 articles in scholarly journals and has written numerous other professional articles. He serves on the editorial review boards of three journals. He has consulted for many leading companies, and much of the material in this book draws on his research and consulting work.

INDEX

Books from Productivity Press

Productivity Press publishes books that empower individuals and companies to achieve excellence in quality, productivity, and the creative involvement of all employees. Through steadfast efforts to support the vision and strategy of con-tinuous improvement, Productivity Press delivers today's leading-edge tools and techniques gathered directly from industry leaders around the world. Call toll-free (800) 394-6868 for our free catalog.

5 Pillars of the Visual Workplace
The Sourcebook for 5S Implementation
Hiroyuki Hirano

In this important sourcebook recently published by Productivity Press, JIT expert Hiroyuki Hirano provides the most vital information available on the visual workplace. He describes the 5'S: seiri, seiton, seiro, seiketsu, shitsuke (which translate as organization, orderliness, cleanliness, standardized cleanup, and discipline). Hirano discusses how the 5S theory fosters efficiency, maintenance, and continuous improvement in all areas of the company, from the plant floor to the sales office. Presented in a thorough, detailed style, *5 Pillars of the Visual Workplace* explains why the 5S's are important and the who, what, where, and how of 5S implementation. This book includes numerous case studies, hundreds of graphic illustrations, and over forty 5S user forms and training materials.
ISBN 1-56327-047-1 / 353 pages, illustrated / $85.00 / Order FIVE-253

20 Keys to Workplace Improvement (revised)
Iwao Kobayashi

The 20 Keys system does more than just bring together twenty of the world's top manufacturing improvement approaches—it integrates these individual methods into a closely interrelated system for revolutionizing every aspect of your manufacturing organization. This revised edition of Kobayashi's bestseller amplifies the synergistic power of raising the levels of all these critical areas simultaneously. The new edition presents upgraded criteria for the five-level scoring system in most of the 20 Keys, supporting your progress toward becoming not only best in your industry but best in the world.
ISBN 1-56327-109-5/ 302 pages / $50.00 / Order 20KREV-B253

Kaizen for Quick Changeover
Going Beyond SMED
Kenichi Sekine and Keisuke Arai

Especially useful for manufacturing managers and engineers, this book describes exactly how to achieve faster changeover. Picking up where Shingo's SMED book left off, you'll learn how to streamline the process even further to reduce changeover time and optimize staffing at the same time.
ISBN 0-915299-38-0 / 315 pages / $75.00 / Order KAIZEN-B253

Non-Stock Production

The Shingo System for Continuous Improvement

Shigeo Shingo

In the ideal production system, information flows from the customer backward through the manufacturing process and results in total elimination of non-value-adding wastes. That means no inventory, inspection, storage, or transportation. Shingo shows that a Non-Stock Production (NSP) system can become a reality for any manufacturer. Find out how, directly from the master himself.
ISBN 0-915299-30-5 / 479 pages / $85.00 / Order NON-B253

One-Piece Flow

Cell Design for Transforming the Production Process

Kenichi Sekine

By reconfiguring your traditional assembly lines into production cells based on one-piece flow, you can drastically reduce your lead time, manpower requirements, and number of defects. Sekine examines the basic principles of process flow building, then offers detailed case studies of how various industries designed unique one-piece flow systems to meet their particular needs.
ISBN 0-915299-33-X / 308 pages / $75.00 / Order 1PIECE-B253

Corporate Diagnosis

Setting the Global Standard for Excellence

Thomas L. Jackson with Constance E. Dyer

All too often, strategic planning neglects an essential first step- and final step-diagnosis of the organization's current state. What's required is a systematic review of the critical factors in organizational learning and growth, factors that require monitoring, measurement, and management to ensure that your company competes successfully. This executive workbook provides a step-by-step method for diagnosing an organization's strategic health and measuring its overall competitiveness against world class standards. With checklists, charts, and detailed explanations, *Corporate Diagnosis* is a practical instruction manual. The pillars of Jackson's diagnostic system are strategy, structure, and capability. Detailed diagnostic questions in each area are provided as guidelines for developing your own self-assessment survey.
ISBN 1-56327-086-2 / 115 pages / $65.00 / Order CDIAG-B253

Cycle Time Management

The Fast Track to Time-Based Productivity Improvement

Patrick Northey and Nigel Southway

As much as 90 percent of the operational activities in a traditional plant are nonessential or pure waste. This book presents a proven methodology for eliminating this waste within 24 to 30 months by measuring productivity in terms of time instead of revenue or people. CTM is a cohesive management strategy that integrates just-in-time (JIT) production, computer integrated manufacturing (CIM), and total quality control (TQC). From this succinct, highly focused book, you'll learn what CTM is, how to implement it, and how to manage it.
ISBN 1-56327-015-3 / 200 pages / $30.00 / Order CYCLE-B253

Productivity Press, Dept. BK, P.O. Box 13390, Portland, OR 97213-0390
Telephone: 1-800-394-6868 Fax: 1-800-394-6286

Implementing a Lean Management System

Thomas L. Jackson with Karen R. Jones

Does your company think and act ahead of technological change, ahead of the customer, and ahead of the competition? Thinking strategically requires a company to face these questions with a clear future image of itself. *Implementing a Lean Management System* lays out a comprehensive management system for aligning the firm's vision of the future with market realities. Based on hoshin management, the Japanese strategic planning method used by top managers for driving TQM through-out an organization, Lean Management is about deploying vision, strategy, and policy to all levels of daily activity. It is an eminently practical methodology emerging out of the implementation of continuous improvement methods and employee involvement. The key tools of this book build on multiskilling, the knowledge of the worker, and an understanding of the role of the new lean manufacturer.
ISBN 1-56327-085-4 / 182 pages / $65.00 / Order ILMS-B253

Kaizen for Quick Changeover

Going Beyond SMED

Kenichi Sekine and Keisuke Arai

Especially useful for manufacturing managers and engineers, this book describes exactly how to achieve faster changeover. Picking up where Shingo's SMED book left off, you'll learn how to streamline the process even further to reduce changeover time and optimize staffing at the same time.
ISBN 0-915299-38-0 / 315 pages / $75.00 / Order KAIZEN-B253

Kanban and Just-In-Time at Toyota

Management Begins at the Workplace

Japan Management Association / Translated by David J. Lu

Toyota's world-renowned success proves that with kanban, the Just-In-Time production system (JIT) makes most other manufacturing practices obsolete. This simple but powerful classic is based on seminars given by JIT creator Taiichi Ohno to introduce Toyota's own supplier companies to JIT. It shows how to implement the world's most efficient production system. A clear and complete introduction.
ISBN 0-915299-48-8 / 211 pages / $40.00 / Order KAN-B253

Poka-Yoke

Improving Product Quality by Preventing Defects

Nikkan Kogyo Shimbun Ltd. and Factory Magazine (ed.)

If your goal is 100 percent zero defects, here is the book for you—a completely illustrated guide to poka-yoke (mistake-proofing) for supervisors and shop-floor workers. Many poka-yoke devices come from line workers and are implemented with the help of engineering staff. The result is better product quality—and greater participation by workers in efforts to improve your processes, your products, and your company as a whole.
ISBN 0-915299-31-3 / 295 pages / $65.00 / Order IPOKA-B253

Visual Control Systems

Nikkan Kogyo Shimbun (ed.)

Every day, progressive companies all over the world are making manufacturing improvements that profoundly impact productivity, quality, and lead time. Case studies of the most innovative visual control systems in Japanese companies have been gathered, translated, and compiled in this notebook. No other source provides more insightful information on recent developments in Japanese manufacturing technology. Plant managers, VPs of operations, and CEOs with little spare time need a concise and timely means of staying informed. Here's a gold mine of ideas for reducing costs and delivery times and improving quality.
ISBN 1-56327-143-5 / 189 pages / $30.00 / Order VCSP-B253

Visual Feedback Photography

Making Your 5S Implementation Click
Adapted from materials by Kenichi Ono

Are you looking for a way to breath some life into your 5S activities in a way that vividly demonstrates your progress? Consider capturing the evolution of your program in photographs. *Visual Feedback Photography* is a simple method for teams to use as they implement workplace improvements, and a means to record changes in the workplace over time. The result is a series of photographs displayed on a workplace chart, providing a clear record of improvement activities related to workplace problem areas.
ISBN 1-56327-090-1 /$150.00 / Order VFPACT-B253

TO ORDER: Write, phone, or fax Productivity Press, Dept. BK, P.O. Box 13390, Portland, OR 97213-0390, phone 1-800-394-6868, fax 1-800-394-6286. Send check or charge to your credit card (American Express, Visa, MasterCard accepted).

U.S. ORDERS: Add $5 shipping for first book, $2 each additional for UPS surface delivery. Add $5 for each AV program containing 1 or 2 tapes; add $12 for each AV program containing 3 or more tapes. We offer attractive quantity discounts for bulk purchases of individual titles; call for more information.

ORDER BY E-MAIL: Order 24 hours a day from anywhere in the world. Our E-mail address is: service@ppress.com. You can also order directly from our on-line catalog, available on the Internet; go to: http://www.ppress.com.

INTERNATIONAL ORDERS: Write, phone, or fax for quote and indicate shipping method desired. For international callers, telephone number is 503-235-0600 and fax number is 503-235-0909. Prepayment in U.S. dollars must accompany your order (checks must be drawn on U.S. banks). When quote is returned with payment, your order will be shipped promptly by the method requested.

NOTE: Prices are in U.S. dollars and are subject to change without notice.

About the Shopfloor Series

Put powerful and proven improvement tools in the hands of your entire workforce!

Progressive shopfloor improvement techniques are imperative for manufacturers who want to stay competitive and to achieve world class excellence. And it's the comprehensive education of all shopfloor workers that ensures full participation and success when implementing new programs. The Shopfloor Series books make practical information accessible to everyone by presenting major concepts and tools in simple, clear language and at a reading level that has been adjusted for operators by skilled instructional designers. One main idea is presented every two to four pages so that the book can be picked up and put down easily. Each chapter begins with an overview and ends with a summary section. Helpful illustrations are used throughout.

Books currently in the Shopfloor Series include:

5S for Operators

5 Pillars of the Visual Workplace

The Productivity Press Development Team

ISBN 1-56327-123-0 / incl. application questions / 133 pages
Order 5SOP-B253/ $25.00

Quick Changeover for Operators

The SMED System

The Productivity Press Development Team

ISBN 1-56327-125-7 / incl. application questions / 93 pages
Order QCOOP-B253 / $25.00

Mistake-Proofing for Operators

The Productivity Press Development Team

ISBN 1-56327-127-3 / 93 pages
Order ZQCOP-B253 / $25.00

TPM for Supervisors

The Productivity Press Development Team

ISBN 1-56327-161-3 / 96 pages
Order TPMSUP-B253/ $25.00

TPM Team Guide

Kunio Shirose

ISBN 1-56327-079-X / 175 pages
Order TGUIDE-B253 / $25.00

TPM for Every Operator

Japan Institute of Plant Maintenance

ISBN 1-56327-080-3 / 136 pages
Order TPMEO-B253 / $25.00

Continue Your Learning with In-House Training and Consulting from the Productivity Consulting Group

The Productivity Consulting Group (PCG) offers a diverse menu of consulting services and training products based on the exciting ideas contained in the books of Productivity Press. Whether you need assistance with long term planning or focused, results-driven training, PCG's experienced professional staff can enhance your pursuit of competitive advantage.

PCG integrates a cutting edge management system with today's leading process improvement tools for rapid, measurable, lasting results. In concert with your management team, PCG will focus on implementing the principles of Value Adding Management, Total Quality Management, Just-In-Time, and Total Productive Maintenance. Each approach is supported by Productivity's wide array of team-based tools: Standardization, One-Piece Flow, Hoshin Planning, Quick Changeover, Mistake-Proofing, Kanban, Problem Solving with CEDAC, Visual Workplace, Visual Office, Autonomous Maintenance, Equipment Effectiveness, Design of Experiments, Quality Function Deployment, Ergonomics, and more. And, based on the continuing research of Productivity Press, PCG expands its offering every year.

Productivity is known for significant improvement on the shopfloor and the bottom line. Through years of repeat business, an expanding and loyal client base continues to recommend Productivity to their colleagues. Contact PCG to learn how we can tailor our services to fit your needs.

Telephone: 1-800-966-5423 (U.S. only) or 1-203-846-3777
Fax: 1-203-846-6883